T0262039

Radio Frequency Identification

Radio Frequency Identification

Edited by **Kevin Merriman**

New Jersey

Published by Clanrye International,
55 Van Reypen Street,
Jersey City, NJ 07306, USA
www.clanryeinternational.com

Radio Frequency Identification
Edited by Kevin Merriman

International Standard Book Number: 978-1-63240-435-0 (Hardback)

Printed in the United States of America.

Contents

Permissions

List of Contributors

Preface

This book has been an outcome of determined endeavour from a group of educationists in the field. The primary objective was to involve a broad spectrum of professionals from diverse cultural background involved in the field for developing new researches. The book not only targets students but also scholars pursuing higher research for further enhancement of the theoretical and practical applications of the subject.

Radio Frequency Identification is a dynamic field with multiple applications. This book includes a comprehensive overview of the techniques of Radio Frequency Identification (RFID) and its applications. RFID based functions create incredible new business opportunities such as the support of autonomous living of aged and disabled persons, resourceful supply chains, and well-organized anti-counterfeiting and improved ecological monitoring. Its data administration, scalable information systems, business procedure reengineering and evaluating reserves are emerging as important technological challenges to applications underpinned by new advancements in RFID technology. It compiles researches from renowned experts on the recent advancements in the RFID field.

It was an honour to edit such a profound book and also a challenging task to compile and examine all the relevant data for accuracy and originality. I wish to acknowledge the efforts of the contributors for submitting such brilliant and diverse chapters in the field and for endlessly working for the completion of the book. Last, but not the least; I thank my family for being a constant source of support in all my research endeavours.

Editor

Commercial Utilization of Mobile RFID

Ela Sibel Bayrak Meydanoğlu and Müge Klein

Additional information is available at the end of the chapter

1. Introduction

The basic functionalities of mobile devices (e.g. phone calls, text messages, browsing the internet) are extended to the interaction with physical objects from the real world as a result of the establishment of mobile devices as ubiquitous and personal computing platforms [1]. In this context, mobile interaction with the physical world, the use of mobile devices as mediators for the interaction with the physical world, become more and more popular [2]. Radio Frequency Identification (RFID) technology is one of the enabling technologies that turn mobile devices – commonly mobile phones – into readers of RFID tags attached to physical objects [3]. RFID technology that is used in the context of physical mobile interaction is mobile RFID. This technology enables mobile devices with embedded micro RFID readers to read RFID tags [4]. Via mobile RFID people can contact RFID tagged objects anywhere [5]. The studies in the relevant literature (see Section 2.1) focus mostly on the realization techniques, architecture of the physical mobile interaction and its perception by users and the consequences of this perception. Advantages gained by physical mobile interaction – realized especially by mobile RFID - have not been discussed comprehensively. This study aims to provide an analysis to address this gap and provide a comprehensive discussion. There are indeed a few studies (see Section 3.2.1) in which advantages of using RFID technology in mobile devices are discussed. However, they are discussed briefly while discussing applications of RFID enabled mobile phones and how to use RFID technology in mobile phones.

This study has the characteristics of a review paper. It investigates the studies about physical mobile interaction, mobile tagging and mobile RFID as well as applications of mobile RFID in the relevant literature with the intention of revealing the competitive advantages gained by using mobile RFID.

The study is organized as follows: The next section provides an overview of different studies concerning the physical mobile interaction, illustrates the techniques and the supporting technologies to realize physical mobile interaction. In section 3, mobile RFID is defined. Afterwards commercial applications in the relevant literature are categorized, in order to define the possible B2C applications enabled by mobile RFID. In section 4, commercial advantages gained by application of mobile RFID are illustrated. Section 5 concludes the study.

2. Physical mobile interaction

2.1. Related work

The explanations below, which are related to the physical mobile interaction and the techniques as well as supporting technologies for the realization of physical mobile interaction, are outlined based on the explanations of the studies about the physical mobile interaction in the literature. The literature review revealed that studies about the advantages of physical mobile interaction – realized especially by mobile RFID - are limited. This limitation provides an opportunity for the execution of this study.

Various points concerned with physical mobile interaction, have been discussed until now. Reference [2] develops a framework called Physical Mobile Interaction Framework (PMIF) and shows an example of the implementation of mobile interaction with the PMIF. Reference [6] describes a generic architecture that supports mobile interaction, discusses techniques for physical mobile interaction and their integration in their architecture. Reference [7] presents an experimental comparison of four physical mobile interaction techniques: touching, pointing, scanning and user-mediated object interaction. It describes the advantages and disadvantages of these techniques based on the executed comparisons. Context-specific preferences for the techniques are also described. These preferences help application designers and developers to decide the integration technique. Based on a user-study, techniques pointing, touching and direct input for mobile interaction are evaluated in the study of reference [1]. In the study of reference [8] the techniques of pointing, scanning and touching are described. Furthermore, several use-cases concerning the techniques are illustrated. In the study, the touching technique is examined closer, and it is described how it can be realized via RFID or NFC (Near Field Communication). Reference [9] investigates user perceptions on mobile interaction with visual and RFID tags and potential usability risks that are due to the limited or erroneous understanding of the interaction technique. Reference [10] presents an analysis, implementation and evaluation of the physical mobile interaction techniques of touching, pointing and scanning. Reference [11] describes a conceptual system that enables the usage of physical posters as gateways to mobile services and a generic architecture for such a system. The services are related to advertisements and information presented on the posters. In the study, two scenarios are described to illustrate the usage of the developed system. Furthermore, places where posters can be found and behavior of people at stops where posters are observable are analyzed. Additionally, expectations of potential users in mobile and context-aware services are described. In order to prove the developed concept, a prototype is also illustrated. Reference [12] investigates mobile interaction with tagged, everyday objects and associated information that is based on the Internet of Things and its technologies. The study focuses on the implementation, design and usability of physical mobile interactions and applications.

2.2. Definition

"Physical mobile interaction (PMI) describes such interaction styles in which the user interacts with a mobile device (e.g. smart phone, PDA) and the mobile device interacts with objects in the real world [7]." PMI enables mobile devices to interact physically with smart objects (tagged objects) and consequently with associated information as well as services [7], [12]. Smart objects can be things, people or locations. Figure 1 visualizes how the physical mobile interaction functions.

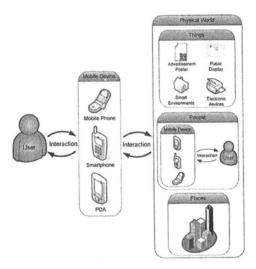

Figure 1. Physical Mobile Interaction [2]

2.3. Techniques for physical mobile interaction

Touching, pointing, scanning and user-mediated object interaction are the techniques that are commonly used for the physical mobile interaction [7], [8]. Based on the following determining factors, application designers and developers select the most appropriate technique to integrate into their applications [7], [10]:

- application context,
- location of the object,
- distance between object and user,

- service related to the object,

- capabilities of the mobile device,

- preferences of the user.

Pointing: By means of this technique, user can select or control a smart object by pointing at it with the mobile device [7]. Users of camera-equipped mobile devices point onto visual markers (e.g. QR-Codes (Quick-Response Codes)) on physical objects. In order to access the stored information on markers, visual markers are interpreted by recognition algorithms [13].

Scanning: According to this technique, mobile device scans the environment for nearby objects. Scanning can be triggered by a user, or the environment is scanned permanently by a mobile device. As a result of scanning, nearby smart objects are listed [7]. User is then free to choose the object with which he wants to connect. After the establishment of the connection, direct input from the user is required [8].

Touching: By means of this technique, user touches a smart object with a mobile device or brings them close together (e.g. 0 to 10 cm) [7], [8]. RFID and NFC are the common technologies for touching interaction [10]. Reference [14] is one of the first to present a prototype for touching interaction via RFID. The prototype uses RFID tags and a RFID reader connected to a tablet computer. It enables an interaction with augmented books, documents and business cards, in order to access links to the corresponding services like ordering a book or picking up an e-mail address [7], [10]. This interaction type is relevant for this study. In this context, it is discussed below how this technique is realized.

User-mediated object interaction: By means of this technique the user types in information provided by the object to establish a link between the object and the mobile device. As user is responsible for the establishment of the link, no special technology is needed for linking. Portable museum guides are good examples for the application of this technique. A visitor using portable museum guide has to type in a number to get information about a desired exhibit or a URL printed on an advertisement poster to get access to the corresponding services [7].

2.4. Supporting technologies for physical mobile interaction

Typical technologies that support physical mobile interactions are RFID, NFC and 2D Barcodes [15], [16].

2D barcodes and QR-codes: A traditional linear (1D/1-dimensional) code contains data in one direction only. 2D barcode is a graphical image that stores information both horizontally and vertically. That is why it can represent more data per unit area than a linear code. Additionally, it can encode several types of data such as symbols, control codes, binary data and multimedia data [15].

Figure 2. QR-Code vs. linear code [15]

Figure 3 includes some examples of 2D codes.

Figure 3. Examples of 2D barcodes [16]

Among barcodes, 2D barcodes are commonly used for mobile applications. QR-Codes were developed by the Japanese Company Denso Wave Corporation in 1994. It is faster than other 2D codes, because it contains three square position patterns that are used for position detection. These patterns are also used to detect the size, the angle and the outer shape of the symbol. When a reader scans a symbol, it first detects these patterns. Once they have been detected, the inside code can be read rapidly by the scanner. Decoding speed of QR-Codes is 20 times faster than that of other 2D codes [15]. These advantages of QR-Codes are the motives for preferring them for mobile applications.

In order to use barcodes for physical mobile interaction, mobile devices have to be equipped with cameras and image recognition algorithms. Using cameras of mobile devices and applying image recognition algorithms, barcodes – thereby products – are identified [3].

RFID: RFID is an Auto-ID technology that enables to identify tagged items by means of radio waves. Main components of a RFID system are:

- *Tag (Transponder):* It consists of an antenna and a microchip. Microchip stores data about the tagged item. Antenna transmits the data about the tagged item to the reader by means of radio waves [17].

- *Reader (Transceiver):* It is a device that communicates with tags through radio waves and reads data on them [18].

- *RFID Middleware:* It is a type of software that is used to consolidate, aggregate, process and filter raw RFID data, which are received from multiple readers, in order to generate useful information for end-users. It transmits also the processed data to backend enterprise applications [19].

- *RFID System Software:* It is software for the communication between tags and readers in order to read tags, write on tags, detect and fix erroneous data as well as to realize authentication for security [20].

- *Backend Enterprise Service:* This service helps to receive filtered RFID data from the middleware and integrate these with existing applications such as ERP, SCM or CRM systems through Application Programming Interfaces (APIs) [19].

NFC: This technology can be seen as an evolution of RFID technology [15]. It is a combination of RFID and interconnection technologies [21]. NFC is compatible with RFID. Both of them use the same working standards and radio frequencies for communication [15]. The differences between these technologies can be listed as follows:

- RFID operates in a long distance range compared to NFC. There is an eavesdropping risk for data exchange. NFC has a short transmission range. That is why NFC-based transactions are inherently secure and there is almost no risk of eavesdropping [8], [15].

- RFID allows only one mobile interaction method, according to which a reader reads or writes a predefined tag. NFC enabled devices allow three different mobile interaction methods. According to the first alternative, NFC enabled mobile device initiates the data transfer by sending a RFID signal to the tag. The tag responds and sends the information it contains back to the mobile device. This type of interaction is congruent with the interaction in RFID systems. According to the second type of interaction, the NFC enabled device acts as a tag (or a smart card). Information on the device can be read by a reader at an interaction point. According to the last type of interaction, direct communication between two NFC enabled mobile devices is possible [8].

NFC has two basic elements: Initiator (called reader in RFID) and target (called tag in RFID). Initiator begins and controls the information exchange. Target responds to the requirements of the initiator. Two modes of operation exist for NFC: active and passive. In the active operation, initiator and target generate their own field of radio frequency to transmit data. In the passive operation, only one of these devices generates the radio frequency field. The other device is used to load modulation for data transfer [21].

3. Mobile RFID

3.1. Definition

Two main ways exist to integrate RFID with a mobile phone, which is a commonly used mobile device for physical mobile interaction: a mobile phone with RFID tags and a mobile phone with a RFID reader [10].

A mobile phone with a RFID tag is a mobile device that includes a RFID chip with some identification information programmed on it. Besides a cell phone antenna used for connection to the network operator, the phone contains a RF antenna for communication with RFID readers. When RF tag equipped phone and reader are within an appropriate range for interaction, the tag information is sent to the reader, and the reader can write some information back to the phone's RFID tag [22].

A mobile phone with a RFID reader is a mobile device that includes a RFID reader. This reader collects data from fixed or mobile RFID tags. The phone also includes an antenna. The phone should have an appropriate reader software for reading and writing tags [22]. The rest of this study focuses on mobile devices that are integrated with RFID readers.

A mobile RFID system works as follows [8], [23]:

- User brings the mobile device equipped with a reader and the object with a tag (smart object) close to each other.

- Reader software in the mobile device activates and decodes tag info, which can be a list of services (e.g. getting more information via an online user manual, changing the state of a smart object such as playing music from the smart phone on your home stereo by simply placing the phone on top of the home stereo) offered by smart object, e-mail address, telephone number, web address, preformatted short message, short text, electronic business card.

- Displaying the decoded info on mobile device.

Below an artificial scenario, that was developed in the context of PERCI-project (PERvasive ServiCE Interaction)[1], is illustrated, in order to highlight how mobile RFID functions. The scenario supports mobile ticketing and payment services. Two posters are used in the scenario that are associated with Web services for mobile ticketing. The first poster allows users to purchase movie tickets for appropriate options like movie title, cinema name, number of tickets and preferred timeslots. The second poster enables to ticket purchases for a public transportation system and offers options like station to start the journey, destination, number of passengers, duration of journey to suggest appropriate tickets. Each option on the posters has a NFC tag and a visual marker. Tags and markers contain or reference the information that the option represents (e.g. name of a cinema) [1], [12]. On the posters action and

---

1 PERCI-project is a project of the collaboration between University of Munich and NTT DoCoMo Euro-Labs [1] and is funded by the latter. The goal of the project is to investigate and develop new methods for mobile interactions with the Internet of Things [24].

parameter tags are used. Action tags contain URLs of different services. Parameter tags provide parameter-values for the invocation of service. In order to determine the service that is to be used (e.g. ordering a movie ticket), an action tag has to be selected first. Then the corresponding parameter tag has to be selected for the invocation of the previously selected service (e.g. movie title or time slot) [1]. Users interact with the posters with their NFC enabled mobile phones that support interaction techniques pointing, touching and direct input [1], [6]. A user can buy a movie and a transportation ticket by pointing and touching his NFC enabled mobile phone on the posters. His mobile phone displays his selections and presents him a payment form. The user enters his credit card details on his phone to proceed and receives an electronic confirmation. The user shows the electronic confirmation on his phone to the transport controller and cinema officer [6].[2]

3.2. Mobile RFID applications

3.2.1. Related work

Applications of mobile RFID span across multiple areas including enterprises, consumer markets, public sector and even private lives. Among the reviewed references, which studied the application possibilities of mobile RFID, some references have a general classification of all possible application areas and some concentrate on a few, special application areas. In order to analyze the references, the reviewed applications are grouped into a classification framework for mobile RFID that considers three main application groups: *Public*, *Business* and *Private* (see Figure 4). *Public* applications include non-commercial applications for public use such as applications for education and health. *Private* applications are also non-commercial applications of RFID based appliances and focus on using RFID in connection with mobile devices in houses or in offices (e.g. RFID tagged food items in smart refrigerators). *Business* applications cover all commercial and non-commercial applications in a business organization such as applications for Supply Chain Management, Customer Relationship Management or Workflow Management.

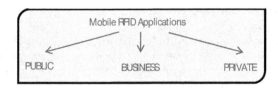

Figure 4. Classification framework for mobile RFID applications

---

2 As mentioned in Section 2.4, NFC technology is an evolution of RFID technology and besides two additional mobile interaction methods, it uses an interaction method that is also used by RFID technology. According to this method, a reader reads or writes a predefined tag. NFC tags of the illustrated scenario of PERCI-project use the mentioned method. In this context, it can be denoted that NFC tags of the scenario do not differ from RFID tags and the scenario of PERCI-project can be used as an example to highlight how mobile RFID functions.

Applications from thirteen references are reviewed in this study, in order to determine the possibilities concerning commercial utilization of mobile RFID. Below, each work is summarized briefly.

Study of reference [15] is one of the few studies that considers directly the use of mobile RFID and proposes four application categories:

1. Applications for education: RFID tags are used to enable learner to access learning content of an object according to the surrounding context.

2. Applications for health: Using mobile RFID for tagged medicines, RFID based patient smartcards, medical RFID patches enable easy access to patient's information and to monitor the health of patients.

3. Applications for entertainment and culture: Mobile RFID is used to enhance visits to museums and art galleries, particularly for guided visits.

4. Commercial applications: Using mobile RFID for any commercial activity such as ticketing, banking or purchasing goods and services.

Reference [25] groups mobile RFID applications into the following zones:

1. Applications for location based services zone: Services related to customer's current location are provided. Service providers deploy RFID tagged items/devices in a location that provide instant real-time information about services available at that location. Downloading bus routes by scanning RFID tagged buses, downloading prices of RFID tagged goods at stores, downloading movie information, trailers, show timings and the nearest theater locations by scanning RFID tagged movie posters, downloading current menu being served at a restaurant by scanning its RFID tag are some examples for applications concerning location based services.

2. Applications for enterprise zone: Mobile RFID applications support company's mobile staff like inventory checkers, field engineers, maintenance and repair staff, and security guards. It supports them in terms of inventory management in real-time, work attendance log, instructions on how to operate tagged items and demonstrating of staff presence at certain locations etc.

3. Applications for private zone: Mobile RFID assists users in their private spaces like home, garden, garage etc. For example, it helps users to make an instant call or send an instant message by scanning RFID tagged photographs and business cards. By scanning RFID tagged household items with a mobile phone, information (e.g. information about the expiration date of milk in the refrigerator or about the last watering time of a RFID tagged plant) can be obtained quickly.

Report of reference [22] focuses mainly on the enterprise market. However, a few consumer applications are presented to show the potential of the technology in the consumer market. Mobile RFID applications are categorized in five groups:

1. Applications for getting real time product information: For example, a service technician touches the machine to service with his smartphone, and up-to-date service information (for example last service date, instructions for additional service) is downloaded at his device.

2. Applications for collecting real-time information: For example, sending specific time and location information about a position or status for some calculations like meter measurement for pricing (measurement is sent for pricing by touching a tag attached on a meter with the mobile device) or like recording travel expenses (a tag attached on the car dashboard sends the starting and ending mileage for an expense report).

3. Applications for automatic asset tracking: Instead of counting devices manually on remote sites, mobile phone can collect info from RFID tags on equipment (PCs, desks, chairs etc.) and send this information to the centralized tracking application.

4. Applications for consumer marketing: Pointing onto a poster enables buying a video, a song etc.

5. Applications to initiate a call: A tag attached on a person's photo can be used to make an automatic call.

Reference [5] analyses commercial applications based on mobile RFID technology and defines three areas for it:

1. Product ordering: RFID tags are used for getting the latest information about products and for ordering them in case of a positive buying decision.

2. Transportation management: RFID tags are used to provide basic information of transported products and to record exception information.

3. Products receiving: RFID tags are used for checking the expected quality of the received products.

Reference [26] examines the utilization of RFID in mobile supply chain management and groups the application areas as follows:

1. Transport and logistics: toll management, tracking of goods.

2. Security and access control: tracking people, controlling access to restricted areas.

3. Supply chain management: item tagging, theft-prevention etc.

4. Medical and pharmaceutical applications: identification and determining the location of staff and patients, asset tracking, counterfeit protection for drugs.

5. Manufacturing and processing: streamlining assembly line processes etc.

6. Agriculture: tracking animals, quality control etc.

7. Public sector: passports, driver's licenses, counterfeit protection for bank notes, library systems etc.

8. Sports and leisure: tracking runners etc.

9. Shopping: facilitating checkout procedures etc.

Reference [27] defines seven scenarios for mobile RFID applications, which are partly inspired by Nokia:

1. Information retrieval: Mobile device helps to receive information on tagged items. Information would be stored in a database, which is accessed via mobile network. For example, a mobile phone user sees an advertisement on a poster and wants to get more information about the advertised product.

2. Data transmission: Means data transfer, for example for reading of electricity meters via a mobile phone.

3. Automated messaging: Messages will be transmitted when the tags are read (e.g. for reporting presence in the office).

4. Voice services: Through tagged items making phone calls simplifies.

5. Device integration: Information retrieved from tags in the environment can indicate to the mobile phone, which could then activate certain functions. For example, when a mobile phone is placed in a car, support for hands-free can be activated or when the mobile phone is in a hospital, it will be blocked.

6. Presence indication: RFID tag on the phone enables readers in the environment to identify the phone. For example, the location data of a person in a building can be used to provide automatic login to a system.

7. Mobile payment: RFID tags in the mobile device store information for payment.

In reference [8] two scenarios for physical interaction are introduced. The first one is a Smart Environment, according to which the user at home can interact with his personal electronic devices. The second one is about Information Heavy Situations, which can be applied for museum visits and guided tours. Transactions in supermarkets and fashion stores, buying car parking tickets, getting tourist information and using active posters are defined as possible utilization of mobile RFID in this study.

In reference [28], literature on mobile commerce (m-commerce) applications are reviewed. The result of this review reveals that location based services, mobile advertising, mobile entertainment services and games, mobile financial applications, product locating and searching, m-commerce in individual companies or industries are the possible m-commerce applications.

Reference [4] defines mobile RFID as a service using mobile devices to download information from RFID tags containing information of a specific area like stores, restaurants and tour sites.

Reference [29] executes a wide study about the utilization of mobile RFID. It describes many applications for RFID (e.g. buying electronic tickets, mobile payment, getting data about products, transportation, stock-trading, services like car rental, bike rental, car parking, taxi ordering, admissions to museums, musical and sports events, automatic call of a technical service hotline) and demonstrates them with case studies.

Using mobile RFID in B2B sector is considered in references [30], [31], [32]. Reference [30] provides information about a B2B case study in the retail industry supported through RFID technology and demonstrates that the RFID network can improve all relevant supply chain processes. In reference [31], mobile RFID technology is used to track and trace a product during supply chain activities (e.g. mobile product authentication service for consumers or an alert service for manufacturers). In reference [32], mobile RFID is used to manage product arrival inspection and loading in the context of transport management.

In Table 1, all applications found in the reviewed references are categorized according to the framework for mobile RFID applications.

Application Areas	Application Types
Public	**Voice Services** [22, 27], **Identity Management** [29], **City Information** (bus routes, train schedules, restaurants, stores) [4, 8, 15, 25, 29], **Mobile Learning** [15], **Health Services and Information** [15, 26], **Mobile Entertainment Services** [25, 28],**Agriculture Management** [26]
Business	**Mobile Commerce**: *Mobile Advertising* [4, 27, 28] via active posters [8, 22, 27, 29], *Product Ordering* [5, 15, 26, 28, 29] (e.g. electronic tickets for buses, car parking, museums, events, [8, 15, 29]), *Service Ordering* (e.g. taxi, technical service, hotel reservation, ski area acess [29], renting a car, bike etc. [29]), *Mobile Payment* [8, 15, 27, 28, 29], **Asset Management** (inventory control) [5, 8, 22,26], **Transportation Management** (arrival inspection, loading, locating, searching, alert service) [5, 26, 29, 30, 31, 32], **Location Based Services** (data transmission for meter readings/pricing, progress reports, work attendance logs) [22, 25, 27, 28, 29]
Private	**Smart Living Environment** (presence indication, device integration, appliance monitoring) [8, 25, 27, 29], **Asset Tracking** [22], **Voice/Messaging Services** [22, 25], **Sports and Leisure** (tracking runners) [26]

Table 1. Mobile RFID applications in categories

3.2.2. Commercial use of mobile RFID

Public and *Private* applications of the mobile RFID classification framework are out of the scope of this study. In order to define commercial applications and the advantages of mobile RFID, this study focuses on *Business* applications. *Business* applications differentiate among *in-house*-applications, *B2B* and *B2C* applications. *In-house*-applications deal with the execution of internal, non-commercial processes in enterprises. *B2B* applications comprise mainly commercial applications in supply chain management with business partners as well as applications for logistic processes. *B2C* processes aim to sell goods to end-consumers. Figure 5 includes an extended classification framework for mobile RFID applications.

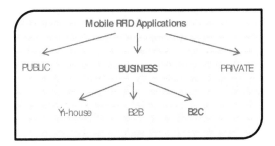

Figure 5. Extended classification framework for mobile RFID applications

As it is seen in Table 1, mobile RFID makes an important contribution to the execution of mobile commerce [5]. M-commerce is a subset of e-commerce and is defined *"as any transaction with monetary value that is conducted via a mobile network"* [33]. Through m-commerce, interaction between supplier and customer is facilitated not only by a mobile network, but also by a mobile customer device. Possible mobile networks for m-commerce are conventional mobile carrier networks, WiFi networks or networks of local frequency technologies for unique identification capabilities for goods (e.g. RFID) (see Figure 6). That is to say, mobile RFID is one of the possible supporting technologies for m-commerce. According to m-commerce supported by mobile RFID, commercial interaction and transaction between suppliers and customers are realized through physical mobile interaction as described above.

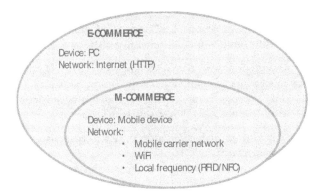

Figure 6. E-commerce vs. M-commerce

This study focuses on B2C m-commerce applications enabled by mobile RFID. As a result of this the m-commerce applications concerning B2C in Table 1 are defined closer:

- **Mobile marketing:** Mobile marketing is a sales approach that helps manufacturers, shopping malls, and service agencies to promote their products and services through interaction with customers via their mobile devices [34]. Kotler [35] defines two basic marketing communications strategies: push and pull strategies. Push-based mobile marketing refers to any content sent by marketers to a mobile device, whether the consumer requests it or not and includes audio, short message service (SMS), e-mail, multimedia messages, or any other pushed advertising content [36]. While push marketing is marketer-initiated, pull marketing is consumer-initiated. Pull-based mobile marketing is defined as any content sent to the mobile consumer upon request [37]. Consumer requests information about products and services that interest him. For classical m-commerce applications enabled by mobile device's Internet or Wifi, both of the mobile marketing communications strategies are applicable. Sending a SMS advertisement to consumer's mobile phone is an example for push mobile marketing. Searching the Internet for a product via a mobile phone is an example for pull mobile marketing. For m-commerce applications enabled by mobile RFID pull marketing strategy is viable. Only if the consumer wants to get more information about a product or a service, or if consumer wants to buy a product or get a service of the RFID tagged item, he can request it. Furthermore, the pull marketing strategy of RFID supported m-commerce is more effective than by classical m-commerce. Because in classical m-commerce, getting information via mobile device's Internet service takes a lot of time and energy of consumer, while through RFID tags information gathering is very quick and convenient [5]. Advertising is an important method for the marketing mix element "communication" (see Section 4). Mobile RFID enables mobile advertising. Through active posters augmented with RFID tags, which advertise a product or a service, or through the tagged items themselves, marketers try to catch consumers' attention. If desired, information about goods is "pulled" very easily via a mobile device. In case of interest, the designated products and/or services can be ordered. Active posters can also be used for location-based mobile advertisements. By using their mobile devices equipped with RFID readers, consumers can read RFID tags, which are placed on boards, and get information about nearby services or products such as restaurants, cinemas [4].

• **Product and service ordering:** Activities concerning product/service ordering do not differ from classical mobile commerce applications. In the context of mobile commerce enabled by mobile RFID, ordering is carried out via a mobile device on Internet or on a product specific network as is done in the classical mobile commerce.

• **Mobile payment:** Payment for product or service ordering occurs also via a mobile device like in classical m-commerce.

4. Advantages gained by mobile RFID

Advantages of using mobile RFID for B2C applications can be grouped as in Figure 7. Mobile RFID is a RFID-based mobile IT application. That is why, the advantages resulted from its characteristics of being an IT system, a mobile solution and a RFID-based technology have to be considered initially.

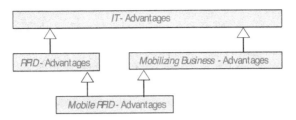

Figure 7. Inheritance classification for mobile RFID advantages

IT-adoption enables organizations to increase efficiency through the automation of processes and information management. As a special IT system, mobile RFID inherits this advantage. RFID technology is a radical innovation in terms of business processes. It does not only automate business processes more efficiently, but it also changes them radically [37]. In addition, it reduces inaccuracy of information caused by transaction errors (e.g. shipment errors, delivery errors etc.) [38], [39]. Substitution of expensive human work through a fully automatic identification system like RFID leads to an increase in information quality, higher data availability and higher speed of process execution, thereby to higher productivity, performance and cost savings [37], [38], [39], [40]. All of these advantages of RFID technology belong also to mobile RFID, which is a RFID-based technology. Mobile business solutions give users the flexibility to operate in a wireless computing environment anywhere [41] and anytime. Users can take advantage of information systems linking business processes among different departments within a company and among companies at remote locations [41]. Ability of accessing a corporate network anywhere and anytime is the primary motive for adopting mobile enterprise solutions [42], [43].

There are also some specific advantages of using mobile RFID in B2C applications. These can be listed as follows:

- **Advantages concerning Marketing Mix:** *"The marketing mix is a combination of tactical marketing tools that a firm uses to satisfy the target market [44]"*. Four Cs marketing mix model, which is adopted in this study,[3] groups the marketing tools into four categories: customer needs and wants, cost to customer, convenience and communication. According to the four Cs marketing mix model firms should sell only products that customers need and want. Consumers are more concerned with total costs of ownership of a product rather than its price. Convenience means the ease of buying and finding the product as well as finding information about the product. Consumers should be provided with the most convenient way possible for purchasing. Under communication, any form of communication (e.g. advertising, public relations, personal selling, viral advertising) between a firm and its consumers is understood [45]. Using mobile RFID has a positive impact on the following elements of four Cs marketing mix model:

---

3 Today's firms tend to execute their activities customer-oriented. Thus, the study adopts customer-focused marketing mix model four Cs instead of product-oriented four Ps marketing mix model.

- *Customer Needs and Wants*: It is essential to offer products that meet customer needs and wants. In order to determine needs and wants of customers, marketers need a good customer database. They may use mobile tags to provide links to specific mobile sites in which through various tools (e.g. questionnaires, voting) information about the needs and wants of customers are collected. The captured information are then analyzed and used to determine offerings for the target customer [46].

- *Convenience*: As mentioned above, ease of finding information about a product is an essential aspect of convenience. Through mobile RFID tags, marketers can provide additional information about their products (e.g. the nutrient content in packaged foods) or events (e.g. concerts, parties, conferences etc.) and facilitate direct downloads (e.g. branded mobile content) [47].

- *Communication*: Advertising is a powerful form of communication and mobile devices are effective communication tools. Consumers use their mobile devices to get tag info that can be an advertisement of a product or a link to a mobile commerce enabled web site. For example, by pointing his/her cell phone onto a poster of a new single, a consumer can get info about the singer, watch the video clip and even buy the song [22]. Mobile RFID enables also location-based mobile advertisements. By using their mobile devices equipped with RFID readers, consumers can read RFID tags, which are placed on boards, and get information about nearby services or products such as restaurants, cinemas [4]. Reference [48] defines using mobile tags within a location-based mobile advertisement publishing system as a convenient way for vendors to create and edit advertisements that include the vendor's location as well as discount coupons stored on a tag.

- **Seamless B2C process:** Mobile RFID enables seamless process flow from advertising to product/service ordering and following to mobile payment with only one mobile device. Thereby, it contributes to solve the media break problem. Off-line products with RFID tags contain information pointers represented as URLs that enable users to access associated on-line contents. For example, a movie poster on a billboard, which is an off-line marketing instrument, can have a RFID tag. This tag enables the user with RFID reader equipped mobile phone to access online information associated with the movie poster (e.g. a short summary about the subject of the movie, comments of movie reviewers), which forms an important online-marketing instrument [49].

- **Ease of information access:** Information access is possible from anywhere to anytime. Information about products or services, which seems interesting for users, can be received immediately via mobile devices [50], [51].

- **Enhanced CRM:** With RFID tagged items and their ordering via mobile devices customer-oriented direct pull marketing strategy is followed. Customers are always able to retrieve valuable product/service information. This customer-oriented characteristic of mobile RFID can increase customer loyalty and lead to repeat purchasing [52].

- **New business models:** Mobile RFID technology leads to new business opportunities for services and products (e.g. the company Flexcar exists only because of its remote vehicle-usage-monitoring system) [22]. Companies that use this technology profile as innovative companies on the market. This is an important competitive advantage for companies [16].

All of the listed advantages have positive impacts on the increase of customer satisfaction, keeping existing customers and thereby increasing customer loyalty as well as on gaining new customers. As customer satisfaction, customer loyalty and an increase in the number of new customers impact directly the revenue of a company, it is not wrong to conclude that mobile RFID has an essential impact on the revenue increase of a company.

5. Conclusion

Although RFID is not a new technology, mobile RFID applications are still in their infancy and their business impact is still unproven. Most of the studies about mobile RFID in the relevant literature are limited to the realization techniques, application possibilities or to case studies. Commercial advantages gained by mobile RFID have not been discussed comprehensively. In this study, based on a literature review B2C applications of mobile RFID are analyzed and commercial advantages of using mobile RFID for B2C applications are illustrated. In this context, first physical mobile interaction concept was defined. Following, mobile RFID was introduced as a supporting technology for physical mobile interaction. After the categorization of mobile RFID applications in the relevant literature, the possible B2C applications enabled by mobile RFID were defined. Finally, commercial advantages of using mobile RFID were illustrated.

This study sheds an insight into the business value of mobile RFID from a commercial viewpoint. Certainly, findings of this theoretical study have to be concretized and validated through case studies in future research.

Author details

Ela Sibel Bayrak Meydanoğlu[1]* and Müge Klein[2]

*Address all correspondence to: meydanoglu@tau.edu.tr

1 Turkish-German University, Faculty of Economics and Administrative Sciences, Department of Business Administration, İstanbul, Turkey

2 Marmara University, Faculty of Administrative Sciences, Department of Business Informatics, İstanbul, Turkey

References

[1] Broll G., Siorpaes S., Rukzio E., Paolucci M., Hamard J., Wagner M., Schmidt A. Comparing Techniques for Mobile Interaction with Objects from the Real World. In:

Proceedings of Pervasive 2007 workshop on Pervasive Mobile Interaction Devices (Permid 2007), 13 May 2007, Toronto, Ontario, Canada; 2007.

[2] Rukzio E, Wetzstein S., Schmidt A. A Framework for Mobile Interactions with the Physical World. http://old.hcilab.org/documents/AFrameworkForMobileInteractions-WithThePhysicalWorld_WPMC2005.pdf (accessed 16 July 2012).

[3] Reischhach F., Michahelles F., Guinard D., Adelmann R., Fleisch E., Schmidt A. An Evaluation of Product Identification Techniques for Mobile Phones. http://www.vs.inf.ethz.ch/publ/papers/dguinard_09_productIdentification.pdf (accessed 17 July 2012).

[4] Ku, J.-E., Lee S. W., Park S. J., Hyun T. H. Mobile RFID Application Service – mRFID service based on Surround Information -. In: Proceedings of the 17th International Regional ITS Conference, 22-24 August 2006, Amsterdam; 2006.

[5] Zhu, W., Wang, D., Sheng H. Mobile RFID Technology for Improving M-Commerce. http://wns.ice.cycu.edu.tw/wireless/Mobile%20RFID/Mobile%20RFID%20technology%20for%20improving%20m-commerce.pdf (accessed 21 July 2012).

[6] Rukzio E., Paolucci M., Schmidt A., Wagner M., Berndt H. Mobile Service Interaction with the Web of Things. In: Proceedings of the 13th International Conference on Telecommunications (ICT 2006), Funchal, Madeira Island, Portugal; 2006.

[7] Rukzio E., Broll G., Leichtenstern K., Schmidt A. Mobile Interaction with the Real World: An Evaluation and Comparison of Physical Mobile Interaction Techniques. In: Procedings of AmI 2007. European Conference on Ambient Intelligence, Darmstadt, Germany; 2007.

[8] Buchmeier F. Mobile Interaction with the Physical World. http://www.lmt.ei.tum.de/courses/hsmt/proceedings/pdf/ws2010/07MobileInteractionPhysical.pdf (accessed 16 July 2012).

[9] Belt S., Greenblatt D., Häkkilä J., Mäkelä K. User Perceptions on Mobile Interaction with Visual and RFID Tags. In: Rukzio E, Paolucci M, Finin T, Wisner P, Payne T (eds.) Proceedings of the Workshop Mobile Interaction with the Real World (MIRW 2006) in Conjuction with the 8th International Conference on Human Computer Interaction with Mobile Devices and Services (Mobile HCI 2006), September 2006; 2006.

[10] Rukzio E., Leichtenstern K., Callaghan V., Holleis P., Schmidt A., Chin J. An Experimental Comparison of Physical Mobile Interaction Techniques: Touching, Pointing and Scanning. http://www.informatik.uni-augsburg.de/de/lehrstuehle/hcm/publications/2006-Ubicomp/ubicomp2006_topoisc.pdf (accessed 17 July 2012).

[11] Rukzio E., Schmidt A., Hußmann H. Physical Posters as Gateways to Context-aware Services for Mobile Devices. http://www.medien.ifi.lmu.de/fileadmin/mimuc/rukzio/PostersAsGateways_WMCSA2004.pdf (accessed 17 July 2012).

[12] Broll G., Paolucci M., Wagner M., Rukzio E., Schmidt A., Hußmann H. Perci: Pervasive Service Interaction with the Internet of Things. http://eprints.lancs.ac.uk/42487/1/ieeeinternet2009.pdf (accessed 17 July 2012).

[13] Broll G., Siorpaes S., Paolucci M., Rukzio E., Hamard J., Wagner M., Schmidt A. Supporting Mobile Interaction through Semantic Service Description Annotation and Automatic Interface Generation. http://eprints.lancs.ac.uk/42334/1/semdesk2006.pdf (accessed 17 July 2012).

[14] Want R., Fishkin K.P., Gujar A., Harrison B.L. Bridging physical and virtual worlds with electronic tags. In: Proceedings of the Conference on Human Factors in Computing Systems (CHI'99), 15-20 May 1999, Pittsburgh, PA; 1999.

[15] Vazquez-Briseno M., Hirata F., Sanchez-Lopez J. D., Jimenez-Garcia, E., Navarro-Cota C., Nieto-Hipolito J.I. Using RFID/NFC and QR-Code in Mobile Phones to Link the Physical and Digital World. http://cdn.intechopen.com/pdfs/31056/InTech-Using_rfid_nfc_and_qr_code_in_mobile_phones_to_link_the_physical_and_the_digital_world.pdf (accessed 17 July 2012).

[16] Erdt Concepts GmbH εt Co. KG. Mobile Tagging – Eine neue Schlüsseltechnologie im Electronic Business auf dem Vormarsch. Fachpublikation der Erdt Concepts GmbH εt Co. KG , Deutschland; 2010.

[17] Kavas A. Radyo Frekans Tanımlama Sistemleri. Elektrik Mühendisliği Dergisi 2007; 430, 74-80.

[18] Karygiannis T., Eydt B., Barber G., Bunn L., Phillips T. Guidelines for Securing Radio Frequency Identification (RFID) Systems - Recommendations of the National Institute of Standards and Technology (NIST), NIST Special Publication 800-98, Computer Security Division Information Technology Laboratory National Institute of Standards and Technology, Gaithersburg; 2007.

[19] Kamoun, F. Rethinking the Business Model with RFID. Communications of the Association for Information Systems (CAIS) 2008; 22 636-658.

[20] Üstündağ, A. RFID ve Tedarik Zinciri. İstanbul: Sistem Yayıncılık; 2008.

[21] Bravo J., Hervás R., Chavira G., Nava S. W., Villarreal V. From Implicit to Touching Interaction: RFID and NFC Approaches. http://mami.uclm.es/nuevomami/publicaciones/HSI-jbravo%20%20%20-G.pdf (accessed 17 July 2012).

[22] Karali D. Integration of RFID and Cellular Technologies. http://www.wireless.ucla.edu/techreports2/UCLA-WINMEC-2004-205-RFID-M2M.pdf (accessed 17 July 2012).

[23] Niklas S. J., Böhm, S. Increasing Using Intention of Mobile Information Services via Mobile Tagging, Proceedings of UBICOMM 2011: The Fifth International Conference on Mobile Ubiquitous Computing, Systems, Services and Technologies, 20-25 November 2011, Lisbon, Portugal; 2011.

[24] PERCI (PERvasive ServiCE Interaction) website. http://old.hcilab.org/projects/perci/ (accessed 17 July 2012).

[25] Konidala D.M., Kim K. Mobile RFID Security Issues. In: Proceedings of SCIS 2006, The 2006 Symposium on Cyrptography and Information Security, January 17-20, Hiroshima, Japan; 2006.

[26] Zarei S. RFID in Mobile Supply Chain Management Usage. IJCST 2010; 1(1) 11-20.

[27] Seidler C. RFID Opportunities for Mobile Telecommunication Services. IT-U Technology Watch, Technical Paper; 2005.

[28] Ngai E.W.T., Gunasekaran A. A review for mobile commerce research and applications. Decision Support Systems 2007; 43 3-15.

[29] Hansen W.-R., Gillert F. RFID für die Optimierung von Gescheaftsprozessen. Munich: Carl Hanser Verlag; 2006.

[30] Wamba S. F., Lefebvre L., Bendavid Y., Lefebvre E. Exploring the impact of RFID technology and the EPC network on mobile B2B eCommerce: A case study in the retail industry. International Journal of Production Economics 2008; 112 614-629.

[31] Kim J., Choi D., Kim I., Kim H. Product Authentication Service of Consumer's Mobile RFID Device. In: IEEE Tenth International Symposium on Consumer Electronics ISCE'2006, June 28-July 1, St. Petersburg, Russia; 2006.

[32] Holmqwist M., Steffanson G. Mobile RFID: A case from Volvo on Innovation in SCM. In: Proceedings of the 39th Hawaii International Conference on System Sciences, January 4-7, 2006, Kauai, Hawaii; 2006.

[33] Clark I. Emerging value propositions for m-commerce. Journal of Business Strategies 2001; 18 (2) 133-148.

[34] Gao A., Küpper J. Emerging technologies for Mobile Commerce. Journal of Theoretical and Applied Electronic Commerce Research 2006; 1 (2) Editorial.

[35] Kotler P., Wong V., Saunders J., Armstrong G. Principles of Marketing. Essex: Pearson Education Limited; 2005.

[36] Mobile Marketing Association. MMA Annual Mobile Marketing Guide: Recognizing Leadership & Innovation; 2006.

[37] Melski A. Grundlagen und betriebswirtschaftliche Anwendungen von RFID. Arbeitsbericht Nr.11. Georg-August-Universitat-Göttingen, Institut für Wirtschaftsinformatik; 2006.

[38] Angels R. Rfid-Technologies: Supply-Chain Applications and Implementation Issues. Information Systems Management 2005; 22: 1 51-65.

[39] Sarac A., Absı N., Dauzere-Peres S. A literature review on the impact of RFID technologies on supply chain management. Working paper. ENSM-SE CMP WP 2009, France; 2009.

[40] Fontanella J. Finding the ROI in RFID. Supply Chain Management Review 2004; 8(1) 13-14.

[41] Eng T-Y. Mobile supply chain management: Challenges for implementation. Technovation 26 2006; 682–686.

[42] Basole R.C. Mobilizing the Enterprise: A conceptual model of transformational value and enterprise readiness. In: 26th ASEM National Conference Proceedings, October 2005, Virginia Beach; 2005.

[43] Scherz M. Mobile Business. Schaffung eines Bewusstseins für mobile Potenziale im Geschaftsprozsskontext. Phd Thesis. Technische Universitat Berlin; 2008.

[44] Smutkupt P., Krait D. , Eisichaikul, V. Mobile Marketing: Implications for Marketing Strategies. International Journal of Mobile Marketing (IJMM) 2010; 5 (2) 126-139.

[45] Lauterborn R. New Marketing Litany: Four P's Passe; C-words take over. Advertising Age 1990; 61 (41) 26.

[46] Bayrak Meydanoğlu E. S. Using Mobile Tagging in Marketing. In: Proceedings of International Conference on IT Applications and Management 2012 (ITAM 8), 28-30 June, İstanbul, Tukey; 2012.

[47] Varnalı K., Toker A. and Yılmaz C. Mobile Marketing, Fundamentals and Strategy. New York et. al.: McGraw Hill Publishing; 2010.

[48] Chyi-Ren D., Yu-Hong L., Liao J., Hao-Wei Y., Wei-Luen K. A Location-based Mobile Advertisement Publishing System for Vendors. In: Proceedings of the Eighth International Conference on Information Technology – New Generations (ITNG), 11-13 April 2011, Las Vegas, Nevada, USA; 2011.

[49] Kim Y-W., Development of Consumer RFID Applications and Services. In: Turcu C. (ed.) Development and Implementation of RFID Technology. Vienna, Austria: InTech; 2009. p497-517.

[50] Curtin J., Kauffmann R.J., Riggins F. Making the 'MOST' out of RFID technology: a research agenda for the study of the adoption, usage and impact of RFID. Information Technology Management 2007; 8 87–110.

[51] Tzeng S.T., Chen W-H., Pai F-Y. Evaluating the business value of RFID: Evidence from five case studies. International Journal of Production Economics 2008; 112 601–613.

[52] Smith A.D. Exploring the inherent benefits of RFID and automated self-serve checkouts in a B2C-environment. International Journal of Business Information Systems 2005; 1(½) 149-181.

Chapter 2

Interferometer Instantaneous Frequency Identifier

M. T. de Melo, B. G. M. de Oliveira,
Ignacio Llamas-Garro and Moises Espinosa-Espinosa

Additional information is available at the end of the chapter

1. Introduction

The rapid development of radar, communication and weapons guidance systems generates an urgent need for microwave receivers to detect possible threats at the earliest stage of a military mission. The microwave receivers used to intercept the RF signals must be able to meet these challenges. Thus, microwave receivers have become an important research area because of their applications to electronic warfare (EW) [1].

The instantaneous frequency measurement (IFM) receiver has been mostly incorporated in advanced EW systems. As to perform the fundamental function, which is to detect threat signals and provide information to the aircrafts, ships, missiles or ground forces, the IFM receiver offers high probability of intercept over wide instantaneous RF bandwidths, high dynamic ranges, moderately good sensitivity, high frequency measurement accuracy, real time frequency measurement and relatively low cost.

IFM started out as a simple technique to extract digital RF carrier frequency over a wide instantaneous bandwidth mainly for pulsed RF inputs. It is been gradually developed to a resourceful system for real time encoding of the RF input frequency, amplitude, pulse width, angle of arrival (AOA) and time of arrival (TOA) for both pulsed and continuous wave (CW) RF inputs. For many electronic support measures (ESM) applications, the carrier frequency is considered to be one of the most important radar parameters, since it is employed in many tasks: sorting, even in dense signal environments; emitter identification and classification; and correlation of similar emitter reports from different stations or over long time intervals, to allow emitter location [2,3].

An IFM receiver is an important component in many signal detection systems. Though numerous improvements have been made to the design of these systems over the years, the basic principle of operation remains relatively unchanged, in that the frequency of an incoming signal is converted into a voltage proportional to the frequency. Microwave interferometers are usually base circuits of the IFM systems. These interferometers most often consist of directional couplers, power combiners/dividers and delay lines [4-8]. As a good example, a coplanar interferometer based on interdigital delay line with different finger lengths, will be presented. Another example of interferometers, but now, implemented with micro strip multi-band-stop filters to obtain signals similar to those supplied by the interferometers was published recently and will be presented here as well [9,10].

2. Important concepts

The system is based on frequency mapping, going from analogical signal into digital words. Any frequency value in the operating band of the system corresponds to a unique digital word. In the process, there is no need to adjust or tune any device. The signal is identified instantaneously. The frequency resolution depends on the longest delay and the number of discriminators.

Let us see how the IFMS maps the incoming signal $x(t)$ into digital words. First of all, consider a sinusoidal signal $x(t) = \sin(\omega t)$ split into two parts, as shown in Fig. 1.

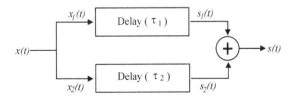

Figure 1. Interferometer used in instantaneous frequency measurement subsystem.

The signals $x_1(t)$ and $x_2(t)$ are then described as

$$x_1(t) = x_2(t) = \frac{\sin(\omega t)}{2} \tag{1}$$

Because of different delays τ_1 and τ_2, one has

$$s_1(t) = x_1(t - \tau_1) \tag{2}$$

and

$$s_2(t) = x_2(t - \tau_2). \tag{3}$$

$S_1(t)$ and $S_2(t)$ are the signals after passing the delay τ_1 and τ_2, respectively. Then the output $s(t)$ is given by the addition of (2) and (3), and after some trigonometric manipulations that sum can be written as

$$s(t) = \sin\left(\frac{2\omega t - \omega(\tau_1 + \tau_2)}{2}\right)\cos\left(\frac{\omega(\tau_2 - \tau_1)}{2}\right). \tag{4}$$

From (4), one can see that the frequency interval between two consecutive maxima or minima of $s(t)$ are given by

$$\Delta f = \left|\frac{1}{\Delta \tau_{2,1}}\right|, \tag{5}$$

where $\Delta \tau_{2,1} = \tau_2 - \tau_1$ is the delay difference between the two branches of the interferometer. Still from (5), it is noticed that from Δf_{max} one gets $\Delta \tau_{min}$ and vice-versa.

As in [1], the frequency resolution is given by

$$f_R = \frac{1}{4\,\Delta\tau_{max}}. \tag{6}$$

A binary code can be generated if

$$\Delta\tau_{max} = 2^{n-1}\Delta\tau_{min}, \tag{7}$$

And this way, the resolution f_r of an n-bits subsystem can be rewritten as

$$f_R = \frac{1}{2^{n+1}\,\Delta\tau_{min}}. \tag{8}$$

Fig. 2 shows the architecture of a traditional instantaneous frequency measurement subsystem (IFMS), where delay lines are used to implement five interferometers as discrimination channels.

Each discriminator provides one bit of the output binary word that is assigned to a certain sub-band of frequency [1]. Wilkinson power dividers are used at the input and output of each interferometer [3]. The output of each discriminator is connected to a detector. The 1 bit A/D converter receives the signal from the amplifier, and attributes "0" or "1" to the output to form the digital word for each frequency sub-band. These values depend on the power level of the received signal. A limiting amplifier is used in IFM input to control the signal gain, to increase sensitivity, and clean up the signal within the band of interest [1], [7].

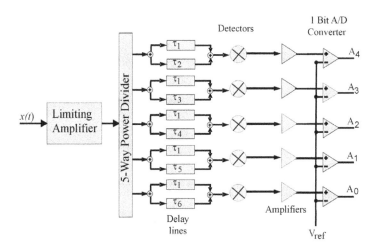

Figure 2. Architecture of a traditional IFM subsystem.

3. Coplanar interdigital delay Line for IFM systems

The schematic drawing of the interdigital delay is shown in figure 3. The particular line consists of 164 interdigital fingers of equal length ℓ, finger width w, finger spacing s and total length L. d is the unit cell length representing the periodicity of the transmission line. If $d \ll \lambda$, an amount of lumped capacitance per unit length C_0/d is added to the shunt capacitance C.

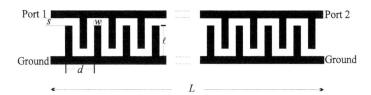

Figure 3. Coplanar interdigital delay line under test.

For the structure shown in figure 3 the phase velocity and the characteristic impedance Z_0, become: $[(C + 2C_0/d) L_S]^{-1/2}$ and $[L_S /(C + 2 C_0/d)]^{1/2}$, respectively. Here, L_S is the series inductance [11]. Due to the fringing electric fields about the fingers, the amount by which the capacitance per unit length increases is greater than the corresponding amount by which the inductance per unit length decreases. In order to exploit the fringing electric fields produced by the fingers, one needs to increase the finger length and keep the finger width fixed.

The $ABCD$ matrix of a lossless transmission line section of length L, line impedance Z_0 and phase constant β is given by

$$\begin{bmatrix} A & B \\ C & D \end{bmatrix} = \begin{bmatrix} \cos(\beta L) & jZ_0\sin(\beta L) \\ (1/Z_0)j\sin(\beta L) & \cos(\beta L) \end{bmatrix} \tag{9}$$

From the above equation one can relate Z_0 to only B and C elements. If we use the conversion from $ABCD$ matrix to S-parameters and assume the source and load reference impedance as Z, we then have [12]

$$Z_0 = \sqrt{\frac{B}{C}} = \left[Z^2 \frac{(1+S_{11})(1+S_{22}) - S_{12}S_{21}}{(1-S_{11})(1-S_{22}) - S_{12}S_{21}} \right]^{1/2} \tag{10}$$

Note that the $ABCD$ matrix is not for a unit cell of the line, it represents the entire transmission line.

Group delay is the measurement of signal transmission time through a test device. It is defined as the derivative of the phase characteristic with respect to frequency. Assuming linear phase change $\phi_{21}(2)\phi_{21}(1)$ over a specified frequency aperture $f(2)f(1)$, the group delay can, in practice, be obtained approximately by

$$\tau g = -\frac{1}{2\pi}\left(\frac{\phi_{21}(2) - \phi_{21}(1)}{f(2) - f(1)} \right) \tag{11}$$

3.1.Intedigitalinterferometer. design and measurement

The structure shown in figure 3 was etched on only one side of an RT/duroid 6010 with relative permittivity $\varepsilon_r = 10.8$, dielectric thickness h = 0.64 mm, conductor thickness t =35μm, w = 0.3 mm, s = 0.3 mm and L = 99 mm. In order to find the line impedance and delay the simulation was carried out varying the finger length ℓ from 0.6 to 4.2 mm and keeping all the other parameters fixed. The devices were fabricated, measured and simulated.

The simulation used sonnet software in order to find the magnitude and phase of the S-parameters, assuming a lossless conductor. Afterwards, equations 10 and 11 were used to find Z_0 and τ_g, respectively. In the experimental procedure each device was connected with coaxial connectors to a HP8720A network analyzer. After carrying out a proper calibration, the devices were then measured. This way, the group delay measurement was implemented, and figure 4 summarizes the group delay results from both measurement and simulation for a frequency range of 0.5-3 GHz. As the finger length increases the lumped capacitance per unit length increases. It slows down the group velocity leading to an increase in the group delay. The longer the finger length, compared to the finger width, the closer it is to a purely capacitive element.

The experimental data of Z_0 were obtained using a reflection measurement in time domain low pass function of the HP8720A. The same devices were all measured again and the results are summarized in figure 5. Looking at the beginning of the curve on the left hand side, the figure 5 seems to agree with the classical coplanar strips formulation, as we found $Z_0 = 99$ Ω for $\ell = 0$ [13]. As we expected, Z_0 decreased as the finger length increased, due to the rise in $2C_0/d$, achieving 50Ω at $\ell = 3.9$mm. As the finger length goes from 0.6 mm to 4.2 mm, τ_g increases about 150% and Z_0 decreases about 45%.

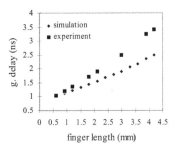

Figure 4. Group delay as a function of finger length at a Frequency range of 0.5-3 GHz.

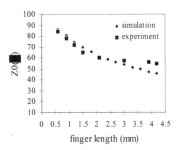

Figure 5. Characteristic Impedance as a function of finger length at a frequency range of 0.5-3 GHz.

These results look promising as far as an IFM application is concerned. Referring to a single stage of a typical IFM, a coplanar unequal output impedance power splitter can be designed to feed two delays with different characteristic impedances. The length of the second delay of each discriminator may be increased to achieve better resolution. The results from figures 4 and 5 may be used together to redesign the coplanar unequal output impedance power splitter to achieve the exact impedance matching. Figure 6 shows a prototype system fabricated based on results of figures 4 and 5. Coplanar wave guide, coplanar strips, coplanar unequal output impedance power splitter and coplanar interdigital delay line are integrated without bends or air bridges. The chip resistors used to increase the isolations between the outputs of the power splitter (and the input of the combiner) are not shown below.

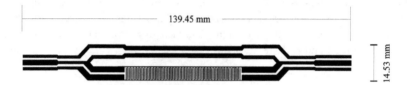

Figure 6. Uniplanar single stage of the IFM under test, scale 1/1

The design has a delay difference of 1.6ns. Two output traces versus frequency from 1.5GHz to 3GHz are presented in figure 7. The theoretical one was obtained using the design equations for a single stage of a typical IFM subsystem [14]. The oscillations in the experimental trace originated from the coaxial connections and the chip resistors bonds.

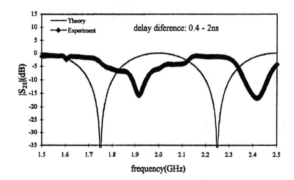

Figure 7. Theoretical interferometer output and measured scattering parameter in dB versus frequency.

4. Interferometer based on band-stop filter for IFM

The IFMS presented now is based on band-stop filter and is shown in Fig. 8. The advantage of using the new architecture is that one has in each channel only multi band-stop filters instead of delay lines and power splitter, as one finds in classical IFMS.

Each word is assigned to only one frequency sub-band to generate a one-step binary code. The response of each multi band-stop filter should be like the one shown in Fig. 9 (a) with discriminators 0, 1, 2, 3 and 4. The discriminator 0 provides the least-significant bit (LSB) and the discriminator 4 provides the most-significant bit (MSB). The form of these responses is suitable to implement the 1 bit A/D converters. Here, let us attribute value 1 if the insertion loss response for the multi band-stop filter is greater than 5 dB, and value 0 for the opposite case. Fig. 9(b) shows the wave form of each 1 bit A/D converter output. According to this example the waveforms at the 1 bit A/D converter outputs are shown in Fig. 9(c). As seen in Fig. 4, this subsystem has its operating band from 2 to 4 GHz, which was divided into 32 sub-bands. Therefore, the resolution obtained was $f_R = 62.5$ MHz.

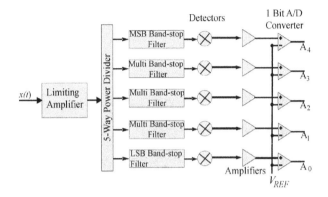

Figure 8. Architecture of an instantaneous frequency measurement subsystem (IFMS) using band-stop filters.

4.1. Multi band-stop filter design and measurement

Rectangular microstrip open loop resonators were chosen to design every discriminator of a five bit IFMS. Frequency response of those resonators presents a narrow rejection band and wide pass band [5] with first spurious out of the working band. Fig. 10(a) shows the top view of a resonator with resonance frequency at 1.9375 GHz. One can see in Fig. 10(b) that the first spurious occurs at 6.140 GHz. Still in this section, it will be shown how this response makes possible the fabrication of a wideband discriminator.

That resonator is placed near to a 50 Ω microstrip transmission line, which was designed with aid of quasi-static analysis and quasi-TEM approximation [8]-[9]. Fig. 11 shows the resonance frequency adjusted by the length $l_1 + l_2 + l_3 + l_4$ of the resonator, which must be approximately half wavelength long [8]. Additionally, there is a coupling gap g given by $l_2 - l_3 - l_4$. Moreover, the coupling distance between the resonator and the main transmission line affects this resonance frequency. This distance also affects the bandwidth of the resonator [8].

Despite the narrow band of the isolated resonators, wide rejection bands are created from coupled arrays. Fig. 12(a) presents 3 sketches of one, two and three resonators, whose resonant frequencies are 2.02, 2.07 and 2.12 GHz, respectively. The line width for the resonators is fixed to be 0.5 mm along this chapter. The ideal coupling distance between resonators is obtained varying $d_{i,j}$ using EM full wave software.

Fig. 12(b) shows the frequency response obtained at ideal coupling distance between them. These distances are chosen to obtain the insertion loss greater than 10 dB over rejection band and also to get this band as large as required. One notices that the coupling between non-adjacent resonators is almost zero. This happens because their resonance frequencies are not very close and the distance between them is large enough. Therefore, the insertion of a new resonator does not change the position of the others already inserted.

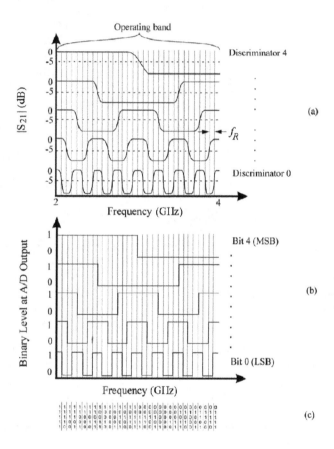

Figure 9. Responses for the IFMS from Fig. 8: (a) desired $|S_{21}|$, (b) A/D converters output, and (c) generated code.

A model of two coupled resonators has been developed by the authors and will be presented in the full chapter.

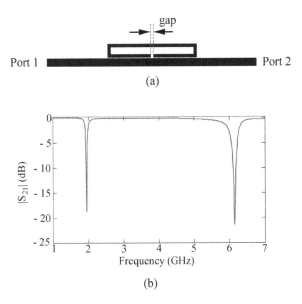

(a)

(b)

Figure 10. (a) Physical structure of a resonator with resonance frequency at 1.9375 GHz, and (b) frequency response of the resonator over a wideband.

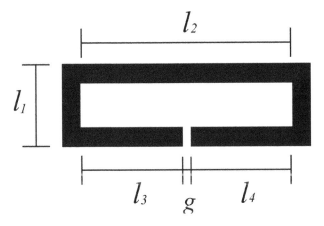

Figure 11. Open loop resonator.

As the desired insertion loss of the discriminator 1 is shown in Fig. 9(a), there must be four rejection bands, where the first one is from 2.125 GHz to 2.375 GHz, regarding the chosen operating band. The resonators are arranged one by one. Fig.13 (a) shows this discriminator with its numbered resonators. The device is designed on a RT6010.2 substrate of relative dielectric constant $\varepsilon_r = 10.2$ and thickness h = 1.27 mm. The 50 Ω transmission line width is 1.2 mm. The gap of every resonator and the distance between the main transmission line and the resonators are kept 0.1 mm for whole structure. Table I shows the coupling distances between the resonators for this device.

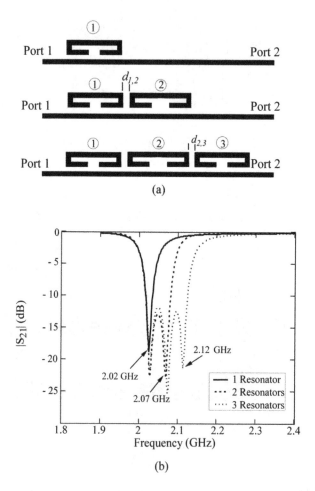

Figure 12. (a) The open loop resonator arrays. The scale has been enhanced for a better comprehension of the devices, and (b) frequency response of 1, 2, and 3 resonators.

Still in Fig. 13(a) one sees four groups of resonators, whose frequency responses and A/D converter outputs are shown in Fig. 15(b). Looking carefully their correlation, Group 1 gives the rejection band over 2 GHz; Group 2 gives the rejection band over 2.5 GHz, and so on. Fig. 13(b) presents the simulated results of the discriminator 1, which agree with the results shown in Fig. 9. One can see the insertion loss level is greater than 10 dB over all rejection bands, and is less than 5 dB over the pass bands. The output A/D converter should generate level zero for $|S_{21}| < -5$ dB and level 1 for $|S_{21}| > -5$ dB. Concerning all the involved $d_{i,j}$, the dimensions of this discriminator are 3 cm wide and 15 cm long. Following the same procedure, the others discriminators are projected, where new resonators configurations will give new desired rejection bands.

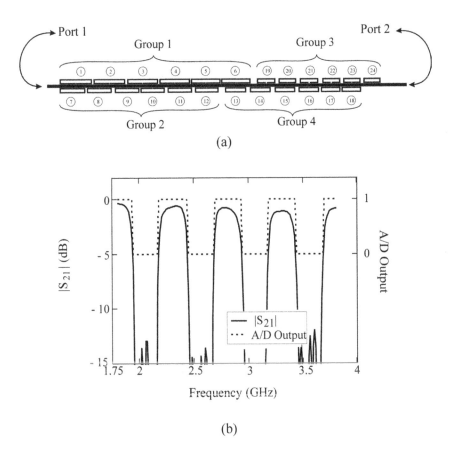

Figure 13. (a) Layout of the discriminator 1, and (b) frequency response of the discriminator 1, and the output of the 1-bit A/D converter; 250 MHz for each rejected band.

Coupling distance between "i" and "j" resonators (mm)	
$d_{1,2} = 0.6$	$d_{13,14} = 1.4$
$d_{2,3} = 0.8$	$d_{14,15} = 1.6$
$d_{3,4} = 0.5$	$d_{15,16} = 1.3$
$d_{4,5} = 0.3$	$d_{16,17} = 0.7$
$d_{5,6} = 0.2$	$d_{17,18} = 0.4$
$d_{7,8} = 0.6$	$d_{19,20} = 1.3$
$d_{8,9} = 1.2$	$d_{20,21} = 1.4$
$d_{9,10} = 0.4$	$d_{21,22} = 1.6$
$d_{10,11} = 1.1$	$d_{22,23} = 1.2$
$d_{11,12} = 1.1$	$d_{23,24} = 1.1$

Table 1. Coupling Distances

The Fig. 14(a)-(e) presents all the projected IFMS discriminators from Fig. 8, having between 23 and 25 resonators. The number of resonators depends on the desired rejection bands. Following the same principle, each group gives only one rejection band, so that discriminators with eight groups have eight rejection bands, as shown in Fig. 14(e). The others, without any specified group, have only one as shown in Fig. 14 (a) and (b). Fig. 15 shows that the simulated and measured results of the five discriminators are in reasonable agreement with each other.

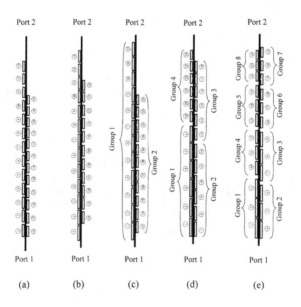

Figure 14. Bandstop filters for implementation of the: (a) discriminator 4 – MSB, (b) discriminator 3, (c) discriminator 2, (d) discriminator 1, and (e) discriminator 0 – LSB.

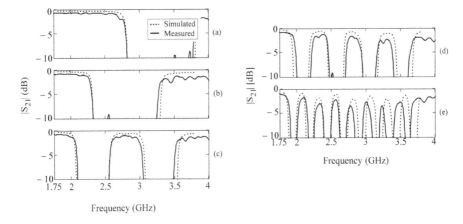

Figure 15. Frequency response of the: (A) Discriminator 4 – MSB, (B) Discriminator 3, (C) Discriminator 2, (D) Discriminator 1, and (E) Discriminator 0 – LSB.

5. Reconfigurable Frequency Measurement (RFM) designs

Fixed IFM designs like the ones discussed in section IV have the advantage of providing instantaneous frequency identification while reconfigurable designs should do a sweep but are very compact in size, making them suitable for portable and handheld systems. RFMs include tuning elements [15] embedded in the designs to produce multibit frequency identification using reconfigurable measurement branches.

An example of RFM architecture is shown in Fig. 16, this design includes a reconfigurable phase shifter used to produce more than one bit. The number of bits will depend on the amount of phase shifts produced by the reconfigurable design; each phase shift will correspond to a specific control voltage in the case of varactors, otherwise switches will be in "on" or "off" state to produce the different phase shifts. The other components shown in Fig. 16 operate in a similar way to the ones exposed in section IV. The RFM can also include reconfigurable bandstop filters [16] instead of the phase shifter to produce a branch that can produce more than one bit as an alternative design.

The switching speed of the tuning elements used in the reconfigurable phase shifter design will mainly determine the detection speed of the subsystem. Solid state components like PIN, varactor diodes, transistors and the use of ferroelectric materials will provide high tuning speeds, (10^{-6} seconds for the PIN and varactor diodes, 10^{-9} seconds for transistors and 10^{-10} seconds for the ferroelectric varactors) while the Micro Electromechanical Systems (MEMS) counterpart will provide slower tuning speeds (10^{-5} seconds) but with the advantage of low power consumption compared with the solid state components. The use of ferroelectric materials results in high tuning speeds with the drawback of having generally high dielectric losses. When designing an RFM it is important to decide which type of technology is adequate for a given application in terms of detection speed, power consumption and device size.

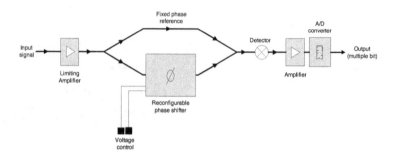

Figure 16. Architecture of a reconfigurable frequency measurement subsystem (RFM) based on phase shifters.

Device size will be mainly determined by the type of technology used to implement the subsystem; the most compact designs can be achieved monolithically, by having the components integrated into a single chip. A monolithic design can include all solid state, MEMS and ferroelectric implementations. Hybrid integrations use microwave laminates or substrates and tuning elements, these include solid state, MEMS and ferroelectric surface mountable components that can be embedded into the design. Hybrid integrations normally involve much larger circuit size compared to the monolithic counterpart, however these components normally involve low cost and simple manufacturing and prototyping techniques.

The most reliable technology is the solid state transistor and the ferroelectric films, followed by the PIN and varactor diode ending with the MEMS components. MEMS packaging can improve device reliability by avoiding contamination or humidity of the movable parts of a switch or varactor. The objective of an RFM is to reduce the size of fixed IFMs by designing branches that can produce more than one bit in the identification subsystem. Size reduction is the main advantage of an RFM over a fixed IFM. A disadvantage over fixed IFMs is that there will be a switching time for the device, so the frequency measurement is not instantaneous.

6. Final considerations

This chapter presented two kinds of interferometers for IFM applications, the first type was a Coplanar Intedigital Interferometer and the second one was based on Multi band-stop filters, which can substitute the interferometers in the IFM Architecture. For the first case, coplanar strips interdigital delay lines were fabricated, simulated and measured at a frequency range of 0.5-3 GHz. As the finger length varied from 0.6 mm to 4.2 mm, keeping all the other parameters fixed, the group delay increased by about 150% and the characteristic impedance decreased about 45%. A prototype of uniplanar IFM with a delay difference of 1.6ns was fabricated and measured based on the results of the characteristic impedance and the group delay.

For the second case, Multi band-stop filters were designed, simulated and measured over a frequency range of 2 GHz. The results show that the use of loop resonators to design the discriminators, instead of delay lines and power splitters, make the simulation and the fabrication easier, as there are no more bends or sloping strips. In addition, one has more control over the resolution, as one can couple the resonators one by one and create the rejection bands. In this process, the association of loop resonators was used to design multi band-stop filters. In light of the above, the use of multi band-stop looks promising as far as planar interferometer identifier is concerned.

The use of loop resonators instead of delay lines and power dividers/combiners, to design IFM systems, decreases the simulating time of the whole structure, as there are no more bends or sloping strips. In addition, one has more control over the resolution, as one can couple the resonators one by one and create the rejection bands. The multi-band-stop filters can substitute interferometers in the IFM system architecture, in a very efficient way. Reconfigurable frequency measurement circuits can considerably reduce the size of the IFMs by using tuning elements embedded into the topologies, resulting in multiple bit circuits by means of reconfigurable frequency measurement branches. RFMs switch between states, thus tuning speed determines the sweep time required for signal detection.

Author details

M. T. de Melo[1]*, B. G. M. de Oliveira[1], Ignacio Llamas-Garro[2] and Moises Espinosa-Espinosa[2]

*Address all correspondence to: marcostdemelo@gmail.com

1 Federal University of Pernambuco, PE, Recife, Brazil

2 Centre Tecnològic de Telecomunicacions de Catalunya (CTTC), Communications Subsystems, Castelldefels, Spain

References

[1] Tsui J. B. Y. Microwave receivers with electronic warfare applications. New York, NY: John Wiley & Sons, 1986.

[2] Weiss A. J. and Friedlander B. Simultaneous Signals in IFM receivers. IEE Proc. Radar, Sonar Navig. 1997, vol. 144, no. 4, August.

[3] Pandolfi C., Fitini, E., Gabrielli G., Megna E. and Zaccaron A. Comparison of Analog IFM and Digital Frequency Measurement Receivers for Electronic Warfare. Proc. 7th European Radar Conference2010, pp. 232-235, October 2010.

[4] Biehl M., Vogt A., Herwig R., Neuhaus M., Crocoll E., Lochschmied R., Scherer T. and Jutzi W. A 4 Bit Instantaneous Frequency Meter at 10 GHz with Coplanar YBCO Delay Lines. IEEE Trans. on Applied Superconductivity 1995, June, vol. 5, no. 2, pp. 2279-2282.

[5] Biehl M, , Crocoll E., Neuhaus M., Scherer T. and Jutzi W. A Superconducting 4 Bit Instantaneous Frequency Meter at 10 GHz with Integrated Resistors and Air-Bridges. Applied Superconductivity 1999, October, vol. 6, nos. 10-12, pp. 547-551.

[6] Wang Y., Su H. T., Huang F. and Lancaster M. J. Wide-Band Superconducting Coplanar Delay Lines, IEEE Trans. on Microwave Theory and Techniques 2005, July, vol. 53, no. 7, pp. 2348-2354.

[7] de Melo M. T., Lancaster M. J. and Hong J. S., Coplanar Strips Interdigital Delay Line for Instantaneous Frequency Measurement Systems, The Institution of Electrical Engineers, London, Digest, reference number: 1996/226, pp. 1/1-1/4, November, 1996.

[8] De Oliveira B. G. M. , Silva F. B. , de Melo M. T. and Novo L. R. G. S. L. A New Coplanar Interferometer for a 5–6 GHz Instantaneous Frequency Measurement System. Proc. 2009 SBMO/IEEE MTT-S International Microwave and Optoelectronics Conference, pp. 591-594, November 2009.

[9] de Souza M. F. A., e Silva F. R. L. and de Melo M. T. A Novel LSB Discriminator for a 5-bit IFM Subsystem Based on Microstrip Band-Stop Filter. Proc. 2008 European Microwave Week, pp. 36-39, October 2008.

[10] de Souza M. F. A., e Silva F. R. L., de Melo M. T. and Novo L. R. G. S. L. Discriminators for Instantaneous Frequency Measurement Subsystem Based on Open Loop Resonators. IEEETransactions on Microwave and Theory and Techniques 2009, September, vol. 57, no. 9, pp. 2224-2231.

[11] Collin R. E. Foundation for Microwave Engineering. McGraw-Hill; 1992.

[12] Kiziloglu K., Dagli N., Matthaei G. L. and Long S. I. Experimental analysis of transmission line parameters in high-speed gas digital circuit interconnects 1991. IEEE Trans., MTT-39, pp. 1361-1367.

[13] Wen C. P. Coplanar waveguide: a surface strip transmission line suitable for nonreciprocal gyromagnetic device applications 1969. IEEE Trans., MTT-17, pp. 1087-1090.

[14] East P. W. Design techniques and performance of digital IFM, IEE Proc., Vol. 129, Pt. F, No. 3, June, 1982.

[15] Minin I., editor. Microwaveand Millimeter Wave Technologies: from Photonic Bandgap Devices to Antenna and Applications. Reconfigurable Microwave Filters, Ignacio Llamas-Garro and Zabdiel Brito-Brito In-Tech, March 2010.

[16] Carles Musoll-Anguiano, Ignacio Llamas-Garro, Zabdiel Brito-Brito, Lluis Pradell, Alonso Corona-Chavez, "Fully Adaptable Bandstop Filter using Varactor Diodes", Microwave and Optical TechnologyLetters, Vol. 52, No. 3, March 2010, pp. 554-558.

Chapter 3

Challenges and Possibilities of RFID in the Forest Industry

Janne Häkli, Antti Sirkka, Kaarle Jaakkola,
Ville Puntanen and Kaj Nummila

Additional information is available at the end of the chapter

1. Introduction

Considerable added value in wood and timber production can be achieved via higher yield and quality of the wood products deriving from improved control of the production processes. The key to improve the production is the identification of the individual wood items in order to utilise exact information of their properties. This can be realized by marking and tracking of tree trunks, logs and sawn wood products to allow the information associated with them to be collected and utilised in all stages of the value chain from forest to the wood product.

Marking and traceability technology for forest industry have been investigated for some time and several technologies have been considered. Recently, UHF RFID technology tailored for the needs of the forest industry has been developed. This Chapter will discuss the unique challenges that the forest industry sets for the radio frequency identification technology and will highlight the benefits and possibilities of the RFID use. Recently developed technical solutions and their trials in production conditions are described.

2. Traceability in the wood supply chain

Individual identification of the wood items (tree, trunk, board, pole, etc.) allows detailed information to be associated with them which can be used to optimise the production. The simplified basic wood supply chain is illustrated in Figure 1.

Figure 1. Simplified general wood supply chain.

First the trees are cut down and then they are transported for processing into products. The processed products are then transported to a secondary manufacturer or to an end-user. The supply chain varies in different countries and for different products as does the level of automation – for example the felling of trees can be done manually with a chain saw or by a forestry harvester. The felling is followed by removal of the branches and in some countries the trunks are cut into logs in the forest by the forestry harvester. The trunks or logs are transported to road side for storage and subsequent transportation to an intermediate storage or directly to a processing plant. The processing steps depend on the product in question – the most common ones being pulp for paper or cardboard making, boards, panels, veneer and poles. Each of these products uses different wood as their raw material and wood with different properties. The highest value round wood in the Nordic countries is used for the production of sawn timber such as boards. This supply chain is discussed in more detail in the following Section.

After processing into the primary product (e.g. boards) the wood products are transported via the associated logistics chain to a secondary manufacturer such as a building component manufacturer or a furniture manufacturer or the end-user (e.g. a consumer or a constructor).

If the wood material could be identified at individual level (trees, trunks, logs, boards, poles, etc.) information can be associated with it – and this information can be traced through the supply chain to optimise the production of wood products.

2.1. Nordic wood supply chain for sawn timber

The Nordic wood supply chain is illustrated in Figure 2.

The trees are felled and cut into logs by a harvester. The harvester also carries out a multitude of measurements on the logs such as measuring their dimensions to determine the volume of wood felled. Next the logs are transported to a pile in the road side by a forwarder. A harvester and a forwarder are shown in Figure 3.

The logs are collected from the road side by a timber truck which transports them either directly to a saw mill or to an intermediate storage. From this storage the logs are transported to a saw mill by truck, train or by floating. At the saw mill the logs are received and sorted into different classes – this sorting is usually based on dimensional measurements using a 3-D laser scanner. In addition to the laser scanner, X-rays may be used to characterise the internal properties of the log. After sorting, the logs of a suitable class are sawn into boards which are then sorted based on their dimensions and quality (e.g. number and size of the knots in them). The sawn boards are usually dried in a kiln and graded for quality after the drying. The graded boards are then stored and packaged for shipping to the end user or to a secondary manufacturer.

Figure 2. Nordic wood supply chain for sawn timber.

Figure 3. Examples of a harvester and a forwarder [1].

2.2. Possibilities and benefits with wood traceability

Currently, the wood material properties are measured when needed in the wood conversion chain and the gathered data is usually lost between the processing steps as illustrated in Figure 4.

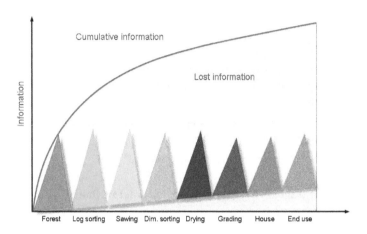

Figure 4. Collection and utilisation of information in the wood supply chain [2].

As the collected information is lost between the processing steps, measurements need to be repeated - such as the measurement of the log dimensions in the forest by the harvester during cutting and the re-measurement of the dimensions in log sorting in the saw mill to determine the volume of the wood for the second time. The lost information also naturally means reduced control over the wood conversion chain from forest to end product as the information related to the wood cannot be traced along the chain.

The traceability of the wood and the associated information can be achieved by identifying the individual wood items – logs and boards, instead of relying on classification of wood and processing in batches of the same class of wood. The benefits of the traceability include:

- Increased quality of the products
- Increased yields
- Reduced production costs.

The quality of the products can be increased with improved control of the production processes by effectively utilising the information collected in the previous stages of the conversion chain. The production process can be optimised based on the individual properties of the wood - processing parameters can be adjusted to better suit the material in question and the most suitable raw material can be used for the product in question.

The yield in the production can be also increased with improved information utilisation and control enabled by the traceability. Downgrading of the boards in the final grading can be reduced when the desired final quality is more consistently achieved. The yield can be also increased by using the right raw material for each product - each type of wood can be used for the most valuable product it is suitable for and less wood material of higher quality is wasted in the production of basic wood products. The production costs can be reduced with improved processing control as the need to 'over-process' the wood is reduced when the actual properties of the wood can be traced instead of relying only on the information on the batch. The improved control over the wood supply and conversion chain together with the more efficient and comprehensive utilisation of the information on the wood material allows also potential new and tailored wood products. Individual identification of the wood items can also be utilised in the logistics – transport planning and control, stock inventory and control in storage, etc.

The traceability in the wood supply chain can also be used to certify the origin of the wood to prevent illegal logging and log theft.

3. Challenges of traceability in the Nordic forest industry

The forest industry presents some unique challenges to the traceability – item marking and identification, and information storage, retrieval and exchange between the different actors in the supply chain. Different supply chains with somewhat different challenges exist for different wood products e.g. pulp and paper, sawn timber, other wood products and energy wood. For simplicity, the discussion is limited here on the sawn timber supply chain with the focus on the round wood.

The wood harvesting takes place in the forest outdoor conditions in rough terrain. The wood is stored outdoors where there is ice, snow, rain, water, dirt, mud, etc. The transportation is by trucks, train or by flotation in bunches from forest to the saw mills. The logs are subjected to impacts with machinery parts, other logs, rocks and the ground. At the saw mill the logs are handled with cranes and conveyors. These conditions are challenging for the log markings and their identification, and for the electronic hardware to be used.

The logs are sawn into boards, which usually destroys the physical markings in the wood and the boards have to be re-marked if full traceability over the chain is targeted. Board marking represents a different challenge from the log marking – the boards are handled in a more controlled industrial environment mostly indoors but the number of boards is larger than that of the logs as each log is sawn in to several boards and the value of each board is lower. Thus the board marking and identification needs to be very inexpensive to be feasible.

In the following Sections, the approaches considered for log marking and identification are discussed together with the specific challenges and limitations related to the use of UHF RFID technology in the forest industry.

3.1. Wood traceability techniques

The traceability can be based on marking and identifying the wood items such as logs and boards. Several methods have been considered for marking tree trunks and logs including painted markings, engraved markings, attached labels with a printed serial number (or other alphanumerical data) or a bar code, fingerprinting techniques based on physical, chemical and/or genetic properties of the wood, and RFID [3]. Markings can be painted or printed on the wood surface and they can be read either by personnel or automatically using machine vision technology (cameras and software). Different coding schemes from simple colour codes to serial numbers and to more advanced codes such as 2D matrix codes have been used. A medium-sized saw mill in Nordic countries typically processes a few million logs per year and thus individual identification of the logs requires a large number of unique codes to be available as the harvested logs may be also transported to several saw mills. Therefore, for unique identification only the more complex codes such as long serial numbers, barcodes or data matrices are feasible. Figure 5 shows examples of a bar code and a matrix code.

(00) 0 0123456 123456789 6

Figure 5. Examples of a GS1-128 code and GS1 DataMatrix [4].

The code markings can be painted, printed, engraved, punched or otherwise imprinted on wood. The main attraction of this kind of markings is the low cost of the marking as each individual marking is inexpensive to make. The most common application of these visual marking codes is printing them onto the boards as the large number of the items and their relatively low value emphasizes the need for low cost marking. The main weakness of the visual markings is their readability – the codes can be obscured by dirt, snow and moisture. A line-of-sight is needed for the camera and optical equipment may need frequent maintenance in dusty industrial environments. Printing of the codes on the surface of wood is also challenging; the surface of wood varies and clear markings are difficult to achieve consistently. For example, markings printed accidentally on dirt or other material on the wood come of when this material comes of the wood. The drying of the wood may distort the shape of the marking and the wood may crack under it.

Multiple techniques have been developed to overcome the problems with visible markings on wood surfaces. One can use attachable labels for smooth printing surface but these labels may also be detached from the wood during the processing steps and the labels can also be covered by dirt, saw dust or other opaque material preventing their reading. Special luminescent inks have been used to improve the readability of the visual markings by increasing the contrast between the markings and the wood surface [5]. Matrix codes allow also for error correction algorithms for improved readability. The achievable identification rate in the board marking with visual codes seems to be in the range of 90-95 % [6].

The use of visual markings on round wood, such as logs and tree trunks, is more complicated than on boards. Logs have to be marked and identified outdoors where the environment is more challenging due to the more frequent occurrence of dirt, water, snow, ice and other materials obscuring the marking. The wood surface also varies more on logs than on sawn boards. In the Indisputable Key –project two log marking methods based on visual markings were tested: luminescent nanoparticle (LNP) ink codes with a handheld marking device and harvester saw integrated printer that sprayed a matrix code or a custom bar code consisting of ink dots or stripes onto the log end [6]. The LNP ink dot or the line markings were read using an infrared camera. The trial achieved 75 % readability of the log markings. The harvester saw integrated marker was tested in marking logs in the forest that we identified using a camera at the log sorting station in a saw mill. Automatic detection rate of the correct identity of the marked logs was nearly 40 % and by eye 74 % of the markings were readable [7].

To overcome the readability problems with visual marking techniques radio frequency identification (RFID) has been tested for log identification. The main advantage of RFID technology is the capability for very high readability – radio waves do not require a line-of-sight and they propagate through most materials excluding highly conductive materials. Thus RFID technology is insensitive to the commonly found dirt, snow, ice and other opaque materials on wood. In the past, RFID trials have used commercial transponders – mainly low frequency (LF, 125 kHz) and high frequency (HF, 13.56 MHz) tags. LF and HF tags have been available much longer and were considered to be better suited for wood marking than UHF (ultra high frequency, ~860-960 MHz) transponders with known problems on moist surfaces e.g. on wet wood and near metal.

Examples of the LF and HF transponder trials are described in [8,9]. In [8] logs were marked by inserting 23 mm long LF tags by Texas Instruments into logs in the forest using a prototype applicator in the harvester. The reading range is reported to have been up to 0.5 m and reading accuracies in the range of 80-90 % were reported at the saw mill. Korten et al [9] report trials with HF transponder cards that were stapled on the logs and read using loop antennas in the forwarder, in a timber truck and at the saw mill. The reported reading range was 0.5-1 m depending on the reading location. Reading is reported to have been reliable.

Reliable automatic log identification is the basis for the traceability in the wood supply chain. UHF RFID technology offers the potential for high readability as the reading range is typically much longer than at LF and HF. Also, a few years ago GS1 introduced standardisation for UHF RFID which facilitates the implementation of the RFID systems. The challenges related to the use of UHF RFID technology in wood supply chain are discussed in the next Section.

3.2. Challenges of UHF RFID technology in wood traceability

Economically viable utilisation of the traceability requires that the wood items can be automatically identified. The item marking should not reduce their value or limit their use as a high quality raw material. The identification of wood items, the marking and reading, should be done without reducing the production efficiency e.g. by slowing it down and the costs related to the traceability should be reasonable to allow the benefits of the traceability to be utilised. The most significant part of the RFID system is the transponder as they are the most numerous component in the system and their performance is the basis of the overall system performance.

Wood is a natural material with varying properties between the trees, logs and boards – and within them. The density of the wood, the grain orientation and the moisture content vary and thus the electromagnetic properties (the complex permittivity) also vary. The varying moisture content has the greatest effect on the permittivity and loss in the wood. The permittivity variation can lead to transponder antenna detuning and the high loss due to the high moisture content attenuates the radio signal. These effects have to be taken into account in the design of the UHF RFID transponder to guarantee a sufficient reading range in all conditions in the wood supply chain. UHF label tags are therefore not suitable for marking fresh wood with high moisture content. In practice, the reading of the tags at saw mills has to be done at distances up to 1 m. The reading range depends mostly on the transponder as the reader operation is governed by the radio regulations defining for example the maximum allowed radiated power.

Reliable identification of the wood items requires that the tags have a high survival rate in the wood processing steps as transponders that have been destroyed or have been detached from the logs or boards cannot be read. In the RFID trials it has been frequently found out that tags glued, stapled or otherwise attached on the logs may be lost in the wood processing – especially during transportation, on conveyors and in debarking. In [9] it is reported that some 75 % of tags attached to the front-end of the log were lost in debarking at the saw mill. In trials carried out by the authors with tags attached onto the surface of the log ends typically up to a few per cent of the transponders were lost in each processing step which results in a significant loss of tags over the supply chain. Therefore, in order to ensure the transponder survival through the whole supply chain the tags has to be inserted inside the wood. Inside the wood the tag is protected from impacts which will improve the transponder survival rate considerable.

The transponder has to be attached on or preferably inserted into the wood by an applicator tool or machine and the tag has to be suitable for reliable and quick application. The application of the transponder should not reduce the production efficiency i.e. the application should not introduce significant delays. The application has to be done automatically where the wood processing is automatic and manual application is possible only if the wood is handled manually, e.g. felled with a chain saw or a reasonably small number of logs are marked. The transponder has to withstand the application to be readable.

In paper making certain materials even in small concentrations are banned from the wood used as the raw material in pulping as they may cause problems in the quality of the paper produced. These materials include most plastics, coal and metal. This represents a challenge in the manufacturing of the transponders as the commonly used materials cannot be used. When round logs are sawn into rectangular boards some of the wood is left over and this wood is commonly chipped and sold to pulp mills. These wood chippings are a high quality raw material for paper making and a valuable by-product for saw mills. As transponders and their pieces may end up into these chippings, the same restrictions on the tag materials apply also to their use in the sawn timber supply. Thus conventional plastics cannot be used in the tags to avoid possible plastics contamination of the wood. The transponder design and materials have to be suitable for inexpensive mass production of the tags as the costs of the transponders typically forms the largest part of a RFID based traceability system.

4. RFID implementation for forest industry

The past RFID trials have focused on using available commercial RFID transponders to mark logs or other wood items. The results of these trials have varied, but in general the transponders intended for other applications have not been optimal for the needs of the forest industry. Therefore, a custom made RFID solution was considered advantageous and was developed in an EU FP6 funded project called Indisputable Key [10]. The following Sections describe the passive UHF RFID solution developed for the supply chain of the forest industry.

4.1. RFID transponder for log marking in sawn timber supply chain

The basis of the traceability utilising RFID is the transponder used to mark the wood items. The requirements for the transponder to be used in log marking in the Nordic sawn timber wood supply chain can be summarised as follows:

- High readability
- Easy attachment into a log
- Harmlessness in pulp and paper making
- Suitability for inexpensive mass production.

These requirements are discussed in Section 3. The required compatibility of the material used with the pulp and paper making processes is perhaps the most constricting requirement for the transponder. Typically a UHF RFID transponder consists of a thin plastic inlay with a metal foil for the antenna to which the microchip is connected and of a hard plastic casing. As common plastics are not accepted in the wood used for pulping, alternative materials were considered. Biopolymers offer an interesting alternative to conventional plastics.

In addition to the chemical compatibility with the paper making processes, the transponder material has to be suitable for insertion into the wood to ensure tag survival in the logs in the wood processing steps in the supply chain. The material has to be mechanically durable; sufficiently hard but not brittle and it may not absorb water. The transponders have to survive several months in the logs. The material should have suitable electromagnetic properties at UHF frequencies – ideally low loss and stable properties. In addition to the suitable chemical, mechanical and electrical properties the material has to be applicable for mass production of the transponders using common plastic fabrication techniques e.g. injection moulding. A suitable bio-composite material meeting these requirements is ARBOFORM® by Tecnaro GmbH [11] and it was selected as the transponder casing material. The ARBO-FORM® material consists of lignin, natural fibres and processing aids. To facilitate the mass production, conventional plastic inlay with aluminium as the antenna pattern material was selected, as the amount of plastic in the inlays ending up into the pulping from the saw mill is negligible. Currently, paper inlays are also available for a non-plastic alternative. The transponder is EPC Class 1 Generation 2 compatible.

The desired high reliability in the wood tracing requires a good survivability of the transponders, which can be only achieved by inserting the tags inside the log. For high readability in the different steps of the supply chain the best location for the tags is in the log end. The transponder size and shape have to be optimised for insertion into the log – several approaches were investigated in the Indisputable Key –project [12] but a wedge-shaped transponder that is punched into the wood was selected [13]. This transponder has the additional advantage of being difficult to remove from the log or to tamper with. The shape of the casing with the inlay inside and the application method are illustrated in Figure 6.

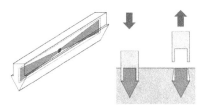

Figure 6. Wedge-shaped transponder and its insertion into the wood.

To achieve good readability in the production conditions on the conveyors, a long reading range is needed. The transponder casing material is somewhat lossy (measured electrical loss tangent is ~0.03 at UHF) which limits the choice of possible transponder antennas to dipole antennas. Wood is a natural material which is not isotropic or homogenous, and the moisture content varies greatly as the wood dries or gets soaked in rain after the tree is felled. The moisture content affects greatly the permittivity and losses of the wood and thus the transponder antenna has to be designed to operate in the wood with varying electromagnetic properties. The moisture content may exceed 100 % of the dry material weight in fresh wood.

The design of the transponder antenna was developed using electromagnetic simulations, laboratory tests and tests in production conditions in saw mills [14]. For electromagnetic simulations, Ansoft HFSS was used. The transponder readability is best when the tag is in the end of the log, as this part is usually exposed in the piles and on conveyor. If the transponder is in the side of the log it may be left under the log or covered by other logs and reading would have to be done through considerable thickness of wood and with the possibility of the tag being pressed against a metal surface. Figure 7 shows the simulator model of the transponder inside the log and the basic layout of the planar dipole antenna inlay.

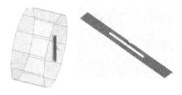

Figure 7. Simulation model of the transponder inside a log and the planar dipole antenna layout.

The planar dipole antenna was optimised for operation inside wet wood with tolerance for varying permittivity caused by varying moisture content in the wood. The electrical properties of wood were measured at UHF and the relative permittivity of the spruce was found to be of the order of 2.3 when fresh and 1.8 after kiln drying. Correspondingly, the loss tangent was 0.08 and 0.03 for fresh and dry spruce. When soaking wet, the relative permittivity of the wood may be even in the magnitude of 10. The final antenna design has the dimensions of 74 mm x 5 mm. Figure 8 shows the reading range measurement in the laboratory together with the measured reading range using TagFormance™ measurement device.

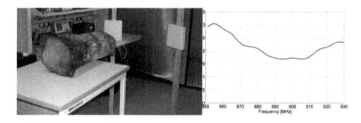

Figure 8. The reading range measurement in the laboratory and the measured reading range.

The reading range from freshly cut wood is approximately 2.5 m at the European UHF RFID frequencies (865.6 - 867.6 MHz) in the laboratory measurements. For inserting the transponder into the log, a simple tool or a manual applicator was developed. The applicator is made from an axe by replacing the blade with a holder for the transponder. Using this applicator, the tag is hit into the end of the log as shown in Figure 9. After some practice an operator may mark up to 100 logs / hour with the first strike success rate of approximately 95 %. In addition to this manual application tool, a prototype for an automatic applicator for a forestry harvester was developed [15].

Figure 9. Application of the tag and the tag inserted into the end of the log.

4.2. RFID readers

Ideally, the traceability in the wood supply chain would reach from the forest all the way to the end user of the wood products - for full coverage of the supply chain RFID transponders would have to be read with RFID readers in every processing step shown in Figure 2. In the Indisputable Key project, the RFID based traceability was used in the round wood supply chain from harvesting in the forest to sawing at the saw mill. RFID readers were used in three processing steps: in the harvesting, at the log sorting and at the sawing where most of the information on the logs is collected and needed – hand-held readers can be used in other processing steps to supplement the fixed readers in the harvester and on the conveyor in the saw mills. The transponders and readers were compatible with the EPC Class 1 Gen 2 air interface standard.

Each reader installation site represents some unique challenges for the RFID reader and for its antennas. The RFID reader has to be able to read the transponders reliably from a practical distance that depends on the location; for example on the conveyor in a saw mill the practical minimum distance from the reader antenna(s) to the transponder in the log is about 1 m as the thickness of the logs varies and sufficient space has to be left to accommodate this variation. The forestry harvester represents the most challenging environment for the RFID readers in the wood supply chain; the reader is subjected to difficult electromagnetic and physical environment in outdoor conditions with rain, snow and ice, vibration, shocks, Nordic four season temperatures and also to occasional impacts. The developed prototype of a vibration and shock resistant RFID reader is described in [14, 15]. The RFID reader features a robust impact resistant IP67 casing, adaptive RF front end for cancellation of reflections from large metal surface in the harvester head and EPC Global Reader Protocol v. 1.1 compatible interface over a CAN-bus to the harvester.

The reader installations at the saw mill were placed in the log sorting where the logs are first received and in the sawing. In these locations the RFID readers are subjected to industrial production conditions – particularly to saw dust and wood splinters, and to the risk of impacts. To protect the readers and to facilitate their installation over the conveyor the commercial readers were enclosed into a robust aluminium casing with the antennas on the outside. The reader used was Sirit Infinity 510 with circularly polarised antennas. Figure 10 shows the reader installations in a saw mill in Sweden in the log sorting station and in the sawing. In the log sorting it was found that antennas in a frame around the conveyor gave more reliable reading of the tags than over the conveyor assembly.

Figure 10. RFID readers in the log sorting and sawing in a saw mill.

RFID readers are used to read the transponders inside the log so that the logs can be identified and information such as measurement data can be associated to the log or the associated data can be retrieved. To identify the individual logs in addition to the reading of the transponder IDs, the ID-code has to be associated with the correct log on the conveyor. In the case of logs in the sawing this is relatively straight-forward as the logs are sawn top first so that the tags in the butt end of the log are always separated by at least the log length. This is based on the automatic applicator always inserting the transponder into the butt end of the log. The speed of the conveyor in sawing is reasonably low as well. In the log sorting the case is more challenging as in some saw mills the logs are not turned before the sorting and the transponders in the log ends can be very close to each other in adjacent log ends on the conveyor. To correctly identify the logs on the conveyor RFID positioning methods such as [16] could be used. In the Indisputable Key projects a simple method based on using the average reading time stamps from several antennas was used to determine the order of the transponders (and logs) on the moving conveyor). When the log separation was larger than about 1 m the logs could be identified reliably but some ambiguity in the log identification remained when the log separation was well below 1 m. The main reason for this was the difficult reading environment in the log sorting shown in Figure 9. There was a flat metal floor under the conveyor that causes reflections; the rapidly changing radio channel causes strong variation in the signal strength and variation of the position where the transponder is read on the conveyor. In the tests in other locations the log identification was significantly more reliable.

4.3. RFID system performance

The RFID system performance in the traceability of round wood in the Nordic wood supply chain was tested in several trials in a saw mill in Sweden and in another saw mill in Finland [15]. In the tests the number of repeated transponder ID readings by the reader was found to be a good indicator of the reading reliability and means to compare reader set-ups. When the tag stays in the field of the reader, the reader keeps reading the ID of the tag repeatedly. With each reading event lasting about one milliseconds, the number of repeated readings indicates how long time the tag has been in the field of the reader. Table 1 shows an example of the observed average number of repeated readings in three tests – in Sweden 164 transponders in 82 logs were run through the log sorting twice, and in a Finnish saw mill 143 test logs with transponders were sawn.

Test	Number of transponders	Reading rate	Number of repeated readings per tag	Standard deviation of the repeats
Log sorting test 1	164	100 %	190	120
Log sorting test 2	164	99.4 %	180	120
Sawing test	143	99.3 %	390	150

Table 1. Reading tests with logs marked with UHF transponders.

Typically the obtained transponder reading rates exceeded 99 % in tests with some 200 logs. In practice, the maximum read rate is 300…600 times per second. As can be seen in Table 4.1 the deviation in the number of repeated ID readings (~120) is rather large compared to the average number of the repeats (180-190) in the log sorting at the saw mill in Sweden, whereas in the other reader location the deviation is smaller in relation to the average number of repeats (150 vs. 390) indicating a more reliable and consistent reading of the transponders. These results also show that for intact normally operating transponders the reading rate can be close to 100 %.

Tests with RFID marked logs were also carried out to determine the log identification rate in the log sorting station in the Swedish saw mill shown in Figure 9 (left-hand side). In this location, reflections from the metal floor caused ambiguity in the reading position on the conveyor and the correct order of logs was unusually difficult to determine. Table 2 summarises the results from three tests where the log marking with RFID tags was carried out both in the forest and in the log yard at the saw mill using the manual applicator or the prototype of an automatic applicator in a forestry harvester.

Log marking	Reader location	Number of read tags in the test	Unique measurement results for the readings	Log identification rate
Automatic in the forest	Log sorting	285	268	94.0 %
Manual in the log yard	Log sorting	218	207	95.0 %
Automatic & manual, all logs for 26 Jan 2010	Log sorting	812	754	92.9 %

Table 2. Examples of identification rates obtained in RFID tests in a Swedish saw mill.

The log identification rate was determined by synchronising the measuring time of the logs by 3D scanner in the log sorting and the RFID tag reading time. Due to the variation in the position of the transponder in the conveyor when it was read by the RFID reader located on the conveyor slightly after the 3D scanner, there was a time window for the time difference the reading timestamp and the 3D scanning timestamp. In the tests, there were also unmarked logs mixed with the RFID marked logs. When there was only one log inside this time window when the RFID tag was read, the log identification was considered successful. The achieved log identification rate was on average about 93 % in the log sorting at this saw mill and in other reading locations the log identification rate was practically the same as the transponder reading rate.

4.4. RFID use in other wood supply chains and processing steps

The promising results in the log identification using UHF RFID in the Nordic round wood supply chain created interest to test the capabilities of the RFID technology in tracing wood in other wood supply chains in the Indisputable Key project. Two other cases were investigated: wooden impregnated poles and sawn timber (boards). Impregnated poles are a product that has a supply chain similar to the round wood supply chain for sawn timber except that the wood used for poles has more stringent requirements and thus a higher value. Additionally, the impregnated wood is not used as a raw material for paper or any other product so there is no limitation for the materials to be used in the RFID transponders. The main challenge in the pole RFID marking is the impregnation process: the poles are impregnated with creosote in high temperatures exceeding +100°C and creosote is a powerful solvent of plastics. The tags are exposed to creosote for an extended time in these high temperatures. The impregnation of the poles destroys most commercial tags as well as the developed biodegradable transponder. After some trials some special materials and high-temperature tolerant commercial tags where found but their high prices made them not feasible for production use. Excluding the destruction of tags in the impregnation, the readability of the UHF transponders in poles was excellent.

The high readability of the RFID tags approaching 100 % caused the desire to try UHF RFID marking of sawn timber, i.e. boards, as the optical marking techniques can typically only reach at best up to 90-95 % readability of the markings in production conditions. The large volumes of the boards sawn and the relative low value of the softwood boards excluded the use of cased transponders (hard tags) due to their price. Thus the only option was to experiment with label tags attached to the boards. The best readability with a label tag on the surface of fresh moist board immediately after sawing was achieved using an inlay indented for near metal applications with good performance in close proximity of detuning materials such as wood – e.g. UPM Raflatac Hammer. The achieved reading range was sufficient for board conveyors to ensure nearly 100 % readability where the reader antenna can be placed approximately 0.4 m away from the boards. However, the application of the label tags on the boards proved to be problematic. Different glues and stapling with plastic staples were tested but the transponder survival on boards in the saw mill in the processing steps from sawing to packing of the dried boards proved to be low - up to 30-40 % of the label tags attached to the boards after the sawing were lost before the packing. Thus the resulting traceability of the boards would be too low for useful applications in the range of some 60 %.

4.5. ICT solution

In the Indisputable Key project an ICT system solution was developed to handle the data storage and transfer to enable efficient utilisation of the collected information by different actors in the value chain.

The ICT System Architecture connects the enterprise business processes to the actual flow of objects. The architecture consist of tags to mark the individual objects, readers to observe the movements of tagged objects, reader data processor to interpret the raw RFID reads to basic observation events and an adapter to create the meaningful business events from the RFID events. The Traceability Services that provides the services to analyse and use the information and the Local ONS that provides the way of publishing the services to the other business partners and the users. Figure 11 presents the overall data flow of the architecture.

The ICT system architecture follows the guidelines set by the EPCGlobal architecture. The EPCglobal Network Architecture Framework is a collection of interrelated standards for hardware, software, and data interfaces, together with core services that are operated by EPCglobal and its delegates, all in service of a common goal of enhancing the supply chain through the use of Electronic Product Codes (EPCs).

Traceability Services Architecture extends the EPCglobal scope by offering the way to use other codes than EPCs and by providing Traceability Services. Traceability Services offers methods to monitor and optimize of the forestry wood supply chain, to research wood property correlations, and of course to trace the wood material throughout the supply chain. By tracing the wood object and processes used to manufacture the wood product the Traceability Services offers the chain-of-custody and environmental product declaration for wood products.

The architecture comprises three modules: Adapter, Collaborative Messaging System and Traceability Services. Adapters are used to acquire traceability information from the processes. The Adapters connect the observations of objects to the process data, generate events and send the events to the Messaging System. The Collaborative Messaging System is responsible for sending the event messages to the right subscribers. The Collaborative Messaging System is also responsible for authentication and authorization. The Traceability Services is responsible for storing the Traceability Data and presenting it to the users in correct format.

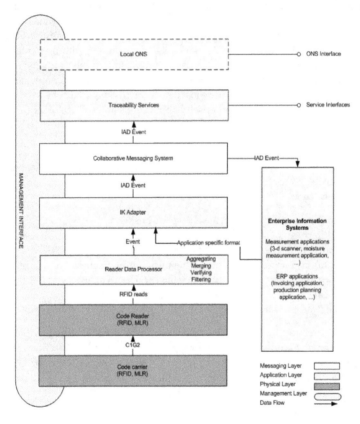

Figure 11. ICT System Architecture.

The interfaces between architecture modules are specified to each message format:

- C1G2, UHF Class 1 Generation 2 Tag Interface standard specifies the interface between RFID readers and RFID tags. The specification describes the interactions between readers and tags and tag operating procedures and commands. The full specification can be found from [18].

- RFID reads are individual reads of a RFID tag. The specification of the protocol used when tag readers interact with upper levels is specified in [19].

- Event is specified to be one observation concerning an individual object. The interface used to transmit the event to the Adapter is EPC global's Filtering & Collection (ALE) Interface, that specifies the delivery of event data to the upper roles. The event in this level could be "At location X in time Y the object with EPC was observed".

- Application specific format is used to connect the business data to the object observations. For example IK Adapter receives a measurements made by 3-D scanner are received as flat-file. The IK Adapter then connects the measurement information to the event information it received from the RFID-reader.

- IAD Event is specified to be one event concerning an individual object.

The Figure 12 presents the architecture when used across enterprises that do not use the same Collaborative Messaging Service. The data flow between enterprises can be realized by using the IAD events. Some application in Enterprise B can subscribe to the events produced in company A. Another connection point is ONS, for example end customers or parties not included into the production chain is to use ONS to look up for the service and use Services provided by the Traceability services to fetch the needed information. For example - customer can fetch a Chain-Of-Custody document or Environmental Product Declaration for object.

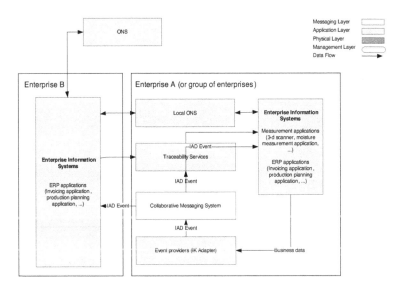

Figure 12. ICT system architecture across enterprises.

The centre of the ICT system architecture is a Collaborative Messaging System that is responsible of transmitting the messages from publishers to correct subscribers. The Publishers are not aware of Subscribers and all the authorization and authentication is performed by the Messaging System.

The Collaborative Messaging System is realized using publish-subscribe pattern. Publish-subscribe is an asynchronous pattern where publishers of events are not sending the events to predefined subscribers. Instead of sending the message to predefined subscriber, message is published with some topic and content. In forestry-wood production system each event must contain event providers ID, detected object ID and a time stamp. Event can also contain some measurement information. For example in log sorting the event can contain measurements that 3-D scanner read from an object.

Any defined Event provider is an event provider in traceability system. IAD event messages are published about events concerning IAD objects and process information events are published about information concerning processes that can't be focused to an individual object. Each IAD event message must contain id of an event provider, id of an object and an observation time, which is the instant of time when the observation took place. An IAD event message can also contain measurement information about an observed object. For example in log reception station a log is measured with a 3-D scanner. These measurements are included into an IAD event message.

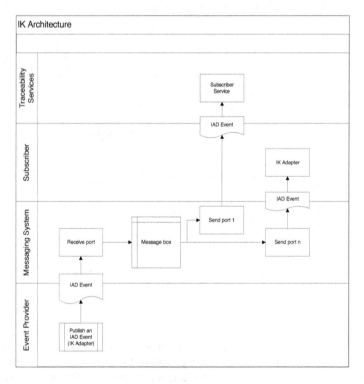

Figure 13. Collaborative Messaging System Data Flow.

Subscription could be topic-based, content-based or a hybrid of these two. In a topic-based subscription a subscriber subscribes for an events published with some topic. In a content based subscription, subscriber receives an event if a content of the event matches to the constraints defined by subscriber. Traceability architecture support hybrid of these two. IAD event providers publish events of a topic and subscribers can define content based subscriptions to one or more topics. For example - as illustrated in Figure 14Example IAD event data flow.

A harvester publishes two events with different topic:

- A LogHarvested event which contains the exact volume, quality and price information about log harvested

- A HarvesterState event which contains information about harvester state (battery, fuel, position, etc...)

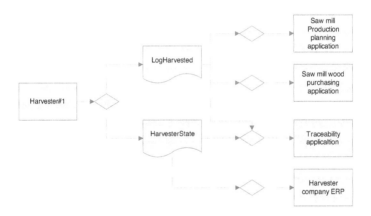

Figure 14. Example IAD event data flow.

There are three different subscribers for the event LogHarvested. Saw mill production planner wants to preplan the production beforehand by knowing the quality and amount of logs that are about to arrive to the saw mill. Saw mill purchaser makes payment based on the log volumes harvested and Traceability application gathers the information for research. For the event HarvesterState there are two subscribers. Traceability application gathers data for research and Harvester company can monitor its harvester status.

By combining information throughout the supply chain the Traceability Services enables new methods of analyzing the wood material. The properties of wood object can be compared between different steps, see Figure 15 Supply chain steps with properties.

For example length in harvesting vs. length in log sorting. Another possibility is to analyze how some property affects some other property. For example, how an area of origin affects the board quality.

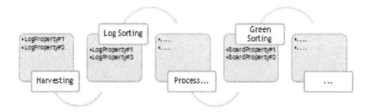

Figure 15. Supply chain steps with properties.

The purpose of Traceability Services is to act as a repository for item level traceability data and process level data and to provide services based on this information. The solution connects the steps of supply chain together and provides a common data model for the whole supply chain. The solution offers services for calculating of Environmental, Economical and Quality KPIs and analysis for the process data that are the basis for the KPI calculations.

5. Discussion

The forest industry represents some unique challenges to the traceability solutions – the data is utilised by different actors in the value chain so that typically the information is produced by one party and the information needs to be utilised by another party that may be outside the supply chain. The basis for the traceability and information utilisation is reliable and affordable identification of the wood. By identifying the wood material and items in the supply chain the associated information can be utilised by different parties. This enables new level of control of the wood conversion chain, tailored and specialised products, and new business models.

The main challenge in enabling the possibilities of the traceability in the forest industry has been the lack of a reliable and inexpensive means to identify automatically the logs and boards in various processing steps along the wood supply chain. The optical marking techniques such as printed markings offer the potential for very low costs but these methods struggle to reach better than 90-95 % success rate in the automatic identification of the wood items in industrial production conditions. The required identification success rate has to significantly exceed 99 % so that the information retrieval becomes a viable option for reacquiring the needed information, e.g. log dimensions. With 90-95 % identification success rate the risk for not being able to retrieve the needed data is some 10 times larger than what is generally considered acceptable – the benefits of the traceability are quickly lost if the information cannot be retrieved for a significant percentage of the wood items.

RFID technology offers the potential for near 100 % success rate in the identification of logs. The main challenge is the cost of the transponders – the acceptable cost for a transponder depends on the value of the wood material in question and on the expected savings and benefits to be obtained through the use of RFID. Currently the acceptable price for RFID varies case by case and there are different opinions on the price level. The price of the tags depends greatly on volumes – large scale mass production lowers the unit price considerably. For a hard tag the price may go as low as a few cents if there is market for sufficiently large volumes – large numbers of tags are needed to push the price down but before the prices are affordable there is not much demand for the tags.

The main challenge in achieving near 100 % identification success rate in RFID based log marking is the application of the tag – the insertion of the transponder into the wood. This has to be done automatically so that the log production efficiency is not significantly reduced by the log marking. Several prototypes of automatic applicators for forestry harvesters have been developed in different research projects but so far no device suitable for long term production use has been successfully built. This is the main technical challenge to be solved before the RFID based log marking can be adopted in large scale in the forest industry. Current solution allows manual tag application for small scale (up to a few thousand logs) log marking, e.g. for marking log batches and piles, or test logs for research and testing purposes, or marking tree trunks or logs when trees are felled manually using chain saws.

6. Conclusion

There are three main types of situations where traceability can be utilized to gain production improvements in forest industry: trouble-shooting, production optimisation and data mining. Trouble-shooting occurs when some end-product or batch deviates from the target quality. With traceability it is possible to trace the defect of quality to its root cause. For example it could be connected to the specific kiln in the saw mill or to a wood batch and its processing history.

Optimisation can be achieved using the traceability information. For example if the spiral grain angle of a log that has been used to produce a board is known, the twist of the board can be estimated. Using this information the board can be placed as a bottom of the drying patch. This can reduce the final twist of these boards by 50%. Traceability information can be used to mine the different correlations between wood properties. For example a window frame producer needs boards with long average distance between knots and wood with this property can be assigned for production of boards for this end product.

The basis of the traceability is reliable identification of wood items to associate and retrieve information on them. To identify the logs in the Nordic round wood supply chain a novel UHF transponder was developed together with robust RFID reader solutions. The novel wedge-shaped transponder is made from pulping compatible materials and it is inserted into the log end. In trials in saw mills the transponder readability was close to 100 % for intact functional tags. An ICT system solution was also developed for the data storage and transfer to utilise the collected information by different actors in the value chain.

The future development of the RFID based traceability should focus on further improving the reliability of the tracing close to 100 % for all logs. The main technical development needed is an automatic applicator suitable for production use in forestry harvesters to achieve high success rate in fast application of tags into logs. For marking high volume lower value items such as boards an inexpensive but sufficiently reliable identification method is needed – currently used printed markings are inexpensive but not highly readable in production conditions. UHF RFID technology has high readability but there are some technical challenges such as the application to the boards to solve – it is also difficult to achieve very low prices for tags if compared to printed markings.

Author details

Janne Häkli[1], Antti Sirkka[2], Kaarle Jaakkola[1], Ville Puntanen[2] and Kaj Nummila[1*]

*Address all correspondence to: kaj.nummila@vtt.fi

1 Sensing and Wireless Devices, VTT Technical Research Centre of Finland, Espoo, Finland

2 Tieto Oyj, Tampere, Finland

References

[1] Product images, Ponsse, http://www.ponsse.com/media-archive/images/products (accessed 3 September 2012).

[2] Usenius, A., Heikkilä A., Usenius T., Future Processing of Wood Raw Material, Proceedings of the 20th International Wood Machining Seminar, June 7 -10, 2011, Skellefteå Sweden, pp. 1-9.

[3] Dykstra, D. P., Kuru, G., Taylor, R., Nussbaum, R., Magrath, W. B., Story, J., Technologies for wood tracking, In: Environmental and Social Development East Asia and Pacific Region Discussion Paper, World Bank, 2002.

[4] GS1. Bar code types: http://www.gs1.org/barcodes/technical/bar_code_types (accessed 3 September 2012).

[5] Juurma, M., Tamre, M., Infrared radiation excited pigment marking technology. In: Proc of 12th International Symp Topical Problems in the Field of Electrical and Power Engineering, Tallinn, Estonia, 2012, pp. 149-150.

[6] Indisputable Key, Final report, Deliverable D1.24, 2010.

[7] Möller, B. Design, development and implementation of a mechatronic log traceability system. Doctoral thesis, Royal Institute of Technology, Stockholm, Sweden, 2011.

[8] Uusijärvi, R. (ed.). Environmentally sustainable and efficient transformation processes (LINESET), SP Technical Research Institute of Sweden, Wood Technology (SP Trätek), QRLT-1999-01476 Final Report, Research report no. P 0309034, 2003.

[9] Korten, S., Kaul, C. Application of RFID (radio frequency identification) in the timber supply chain, Croatian Journal of Forest Engineering, Vol. 29, No. 1, 2008, pp. 85-94.

[10] Indisputable Key project, website, http://www.indisputablekey.com (accessed 3 September 2012).

[11] Tecnaro GmbH, Arboform, http://www.tecnaro.de/english/arboform.htm (accessed 3 September 2012).

[12] Indisputable Key, Forest RFID Transponder and Reader Design, Deliverable D4.10, 2009.

[13] Häkli, J., Jaakkola, K., Nummila K., Saari J-M, Axelsson B., Kolppo K. Transponder, Transponder Kit, Method Of Applying The Transponder And Product Comprising The Transponder, International patent application number 20110315777

[14] Häkli, J., Jaakkola, K., Pursula, P., Huusko, M., Nummila K. UHF RFID based tracking of logs in the forest industry, 2010 IEEE International Conference on RFID (IEEE RFID 2010). Orlando, FL, USA, 14-16 April 2010, pp. 245-251.

[15] Indisputable Key, Forest RFID System Operation, Deliverable D4.11, 2010.

[16] Viikari, V., Pursula, P., Jaakkola, K. Ranging of UHF RFID Tag Using Stepped Frequency Read-Out, Sensors Journal, IEEE, Volume: 10, Issue: 9; Sept. 2010 Page(s): 1535 - 1539

[17] The EPCglobal Network Architecture Framework Version 1, July 2005. Available on the EPCglobal website at: http://www.epcglobalinc.org/standards_technology/Finalepcglobal-arch-20050701.pdf (accessed 3 September 2012).

[18] EPCglobal, "EPC™ Radio-Frequency Identity Protocols Class-1 Generation-2 UHF RFID Protocol for Communications at 860 MHz – 960 MHz Version 1.0.9," EPCglobal Standard, January 2006, http://www.epcglobalinc.org/standards/uhfc1g2/ uhfc1g2_1_0_9-standard-20050126.pdf. (accessed 3 September 2012)

[19] EPCglobal, "EPC™, Reader Protocol Standard, Version 1.1, http://www.epcglobalinc.org/standards/rp/rp_1_1-standard-20060621.pdf (accessed 3 September 2012)

[20] EPCglobal, "Application Level Events (ALE) Standard" February 2008, ALE 1.1 Standard - Part 1 http://www.epcglobalinc.org/standards/ale/ale_1_1-standardcore-20080227.pdf - ALE 1.1 Standard - Part 2 http://www.epcglobalinc.org/standards/ale/ale_1_1-standard-XMLandSOAPbindings-20080227.pdf (accessed 3 September 2012)

[21] EPCglobal - ONS Standard v. 1.0 - http://www.epcglobalinc.org/standards/ons/ ons_1_0-standard-20051004.pdf (accessed 3 September 2012)

Chapter 4

Application of Mobile RFID-Based Safety Inspection Management at Construction Jobsite

Yu-Cheng Lin, Yu-Chih Su, Nan-Hai Lo,
Weng-Fong Cheung and Yen-Pei Chen

Additional information is available at the end of the chapter

1. Introduction

Jobsite safety management is very important subject special in construction management. Managing jobsite safety-related inspection information plays an important role in the view of safety management. Managing jobsite safety management effectively is extremely difficult owing to various participants and environments in construction jobsite. With the advent of the Internet, web-based information management solutions enable information dissemination and information sharing at the construction jobsite. Generally, jobsite safety-related managers and engineers require access to the important check locations to handle inspection work in construction jobsite. However, jobsite safety-related managers and engineers generally use sheets of paper to handle various types of inspection checklists and entry check record. Consequently, there is serious rework progress regarding the data capture and entry in inspection progress. Furthermore, current desktops and notebooks are not suitable for inspection work at the jobsite because of problems in transportability.

In order to solve the above problems, this study presents a novel system called Mobile RFID-based Safety Inspection Management (RFIDSIM) system for jobsite safety management and providing safety inspection information sharing platform among all participants using web technology and RFID technology. The RFIDSIM is then applied in a construction commerce project jobsite in Taiwan to verify our proposed methodology and demonstrate the effectiveness of safety inspection progress in construction jobsite. The combined results demonstrate that, a RFIDSIM system can be a useful web safety inspection management platform by utilizing the RFID approach and web technology.

Integrating Near field communication (NFC) technology and mobile devices such as NFC Smartphone, Radio Frequency Identification (RFID) scanning and data entry mechanisms, can help improve the effectiveness and convenience of information flow in the safety inspection management. The combined results demonstrate that, an RFIDSIM system can be a useful mobile RFID-based jobsite safety management platform by utilizing the NFC and web technologies. With appropriate modifications, the RFIDSIM system can be utilized at any jobsite inspection and management progress for jobsite safety management divisions or suppliers in support of the RFIDSIM system.

2. Problem statement

Jobsite safety management performance can be enhanced by using RFID technology for information sharing and communication. There are many jobsite safety checkpoints locations need for tracked and inspected for jobsite safety management. Information acquisition problems in inspection management follow from information being gathered from jobsite safety checkpoints locations. The effectiveness of information and data acquisition influences the efficiency of jobsite inspection management. Usually, project managers and safety staff members generally use sheets of paper and/or field notes for jobsite safety inspection progress in Taiwan construction jobsite. Restated, existing means of processing information and accumulating data are not only time-consuming and ineffective, but also compromise jobsite safety management in information acquisition. Such means of communicating information between jobsite safety checkpoints locations and jobsite office, and among all participants, are ineffective and inconvenient. The primary problems in inspection regarding to data capture and sharing based on experts interviews are as follows: (1) the efficiency and quality are low, especially in the safety inspection progress in construction jobsite through document-based media, (2) there are serious rework progress regarding the data capture and input in safety inspection progress, and (3) there are serious problems regarding to inspection information collection and responding during safety inspection progress. However, few suitable platforms are developed to assist jobsite office with capturing and sharing the jobsite safety inspection information when jobsite office needs to handle inspection information and inspection management work. Therefore, to capture data effective and enhance inspection information collection and respond in construction jobsite will be primary and significant challenge in the study.

3. Research objectives

This study utilizes the RFID and web technology to enhance the inspection management progress and effectiveness in jobsite safety management. This system is controlled by the management division, and provides project managers and safety staff members with real-time checkpoints-related information-sharing services, enabling them to dynamically respond to the entire jobsite safety management network. This study develops Mobile RFID-based Safety Inspection Management (RFIDSIM) system to improve efficiency and cost-effectiveness of jobsite safety management, improve practical communication among participants, and increase flexibility in terms of service delivery and response times. RFID-SIM system is a web-based system for effectively integrating managers, safety staff members and relative members, to enhance the jobsite safety management in the construction. Utilizing smartphone with NFC technologies can extend RFIDSIM systems from offices to jobsite safety checkpoints locations. Data collection efficiency can also be enhanced using RFID-enabled smartphone with NFC technology to enter and edit data on the jobsite safety checkpoints locations. By using web technology and RFID-enabled smartphone with NFC technologies, the RFIDSIM system for the management division has tremendous potential to increase the efficiency and effectiveness of information respond and management, thus streamlining services jobsite safety management processes with other participants. The portal and RFID-enabled smartphone with NFC technology enable safety staff members to update data from the jobsite safety checkpoints locations and immediately upload it to the system; project managers can receive inspection information and make better decisions regarding future jobsite safety management and control. The main purposes of this study include (1) developing a framework for a mobile jobsite safety inspection system; (2) applying such a system that integrates RFID-enabled smartphone with NFC technology to increase the efficiency of safety inspection data collection in the jobsite, and (3) designing a web-based portal for jobsite safety management and control, providing real-time information and wireless communication between jobsite office and jobsite safety checkpoints locations. Figure 1 illustrates solutions used in a polite test utilized RFIDSIM system in Taiwan construction jobsite. With appropriate modifications, the RFIDSIM system can be utilized at any jobsite safety inspection and management application in construction.

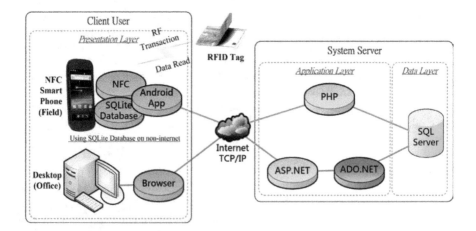

Figure 1. RFIDSIM System Framework Overview

4. Background research

RFID is an automatic identification solution that streamlines identification and data acquisition, operating similarly to bar codes. Automatic identification procedures have recently become very popular in numerous service industries for purchasing and distribution logistics, and in manufacturing companies and material flow systems. Jaselskis and Anderson (1995) investigated the applications and limitations of RFID technology in the construction industry, and attached read/write RFID tags to the surfaces of concrete test that were cast from the job site to test lab. This RFID technology has been widely applied in many areas in the construction industries for the following reasons: (1) to provide owners and contractors with information to enhance operation using RFID technology (Jaselskis and Tarek, 2003); (2) to propose a novel concept of "parts and packets unified architecture" in order to handle data or information related to a product carried by product itself by utilizing RFID technology (Yagi et al., 2005); (3) to apply RFID technology as a solution to problems in pipe spools, and identify potential economic benefits from adopting RFID technology in automated tracking (Song et al., 2006); (4) to apply RFID combined with GIS technology in order to locate precast concrete components with minimal worker input in the storage yard (Ergen et al., 2006); (5) to improve the efficiency of tracing tools and tool availability using RFID (Goodrum et al., 2006); (6) to develop mobile construction supply chain system integrated with RFID technology (Wang et al., 2006); (7) to describe a prototype of an advanced tower crane equipped with wireless video control and RFID technology (Lee et al., 2006); (8) to improve tracing of material on construction using materials tagged with RFID tags (Song et al., 2006); (9) to present strategy and information system to manage the progress control of structural steel works using RFID and 4D CAD (Chin et al., 2008); (10) to enhance precast production management system integrated with RFID application (Yin et al., 2009), and (11) to present a new methodology for managing construction document information using RFID-based semantic contexts (Elghamrawy and Boukamp, 2010).

The use of technology to improve delivery process control is not a novel concept. Many industries have applied barcodes to track materials for many years. Construction companies began to examine the use of barcodes for tool management in the early 1990s. Although barcode is an established and affordable technology, it has presented problems in the construction industry due to the short read range and poor durability of barcodes — a barcode requires a line of sight, and becomes unreadable when scratched or dirty.

An RFID system is composed of an RFID tag and an RFID reader. The RFID tag comprises a small microchip and an antenna. Data are stored in the tag, generally as a unique serial number. The RFID tags can be either passive (no battery) or active (battery present). Active tags are more expensive than passive tags and have a read range of 10–100 meters. Passive tags have a read range of 10mm to approximately 5m (Manish and Shahram, 2005). The vast majority of RFID tags applied in the construction industries are passive.

The RFID reader functions as a transmitter/receiver. The reader transmits an electromagnetic field that "wakes up" the tag and provides the power required for it to operate (Lahiri, 2005). The tag then transfers data to the reader via the antenna. This data are then read by the RFID reader, and transferred to a Pocket PC or computer. Unlike barcodes, RFID tags do not require line-of-sight to be read; they only need to be within the reader's radio range. Additionally, RFID tags can be read through most materials. RFID tags are shrinking, with some measuring only 0.33mm across. Although RFID systems can apply different frequencies, the most common frequencies are low (125KHz), high (13.56MHz) and ultra-high (UHF) (850–900MHz) (Lahiri, 2005).

RFID (Radio Frequency Identification) is a tagging technology that is gaining widespread attention due to the great number of advantages that it offers compared to the current tagging technologies being used today; like barcodes. Near Field Communication, or more commonly known as NFC, is a subset of RFID that limits the range of communication to within 10 centimetres or 4 inches. Compared with Bluetooth and infrared, the main characteristic of NFC is quick, easy, security. Although the NFC data transmission speed is far less than the Bluetooth and infrared, the device only by the unilateral power supply to the operation of the device near to the rapid induction features, and greatly enhance the ease of use. Furthermore of the NFC device requires very short communication distance. NFC technologies help to improve data security because it can reduce the data to the risk of being intercepted or stolen.

In recent years, due to the rapid development and popularization of the smartphone, a growing number of mobile phone NFC functionality into the standard features. The NFC-enabled mobile phone through the sensor to read high-frequency band RFID tags or other NFC-enabled devices for data transmission via a simple touch can, and are widely used in the identification, communication, information obtained, consumption and other purposes provide fast and convenient communication, this study using NFC phones used in patrol operations and identification of the site staff can enhance the operating convenience of mobility and reduce the cost of equipment to build, in order to improve the job the best solution.

Notably, RFID systems are one of the most anticipated technologies that will potentially transform processes in the engineering and construction industries. In the construction industry, RFID technology can be utilized with smartphone, thereby allowing staff members to integrate seamlessly safety work processes in the jobsite, due to the ability to capture and carry data. With a NFC technology plugged into a smartphone, the RFID-enabled smartphone is a powerful portable data collection tool. Additionally, RFID readings increase the accuracy and speed of information communication, indirectly enhancing performance and productivity.

The advantages of using mobile devices in the construction industry are well documented (Baldwin et al., 1994; Fayek et al., 1998; McCullough, 1997). Moreover, mobile devices have been applied in numerous construction industries, to provide the following support: (1) providing wearable field inspection systems (Sunkpho and Garrett, 2003); (2) supporting pen-based computer data acquisition for recording construction surveys (Elzarka and Bell, 1997); (3) supporting collaborative and information-sharing platforms (Pena-Mora and Dwivedi, 2002); (4) using mobile computers to capture data for piling work (Ward et al., 2003), and (5) utilizing mobile devices in construction supply chain management systems (Tserng et al., 2005).

5. System implementation

The RFIDSIM system has three main components, a smartphone, RFID and a portal. Significantly, both the smartphone and RFID components are located on the client side, while the portal is on the server side. All inspection-related information acquired by safety staff members within the RFIDSIM system is recorded in a centralized RFIDSIM system database. All staff members can access required information via the portal based on their access privileges. The RFIDSIM system extends the jobsite safety management from the jobsite office to safety checkpoint locations to assist with safety inspection services, while the RFIDSIM system primarily deals with data transactions in all departments or systems integration. The RFIDSIM system consists of a mobile inspection management portal integrated with RFID-enabled smartphone and RFID technology. Each module is briefly described below.

5.1. RFID subsystem of RFIDSIM system

The RFID technology can be either a passive or active system. The major difference between an active and a passive RFID system is that an active tag contains a battery, and can transmit information to the reader without the reader generating an electromagnetic field. The case study uses HF passive RFID technology due to budget restrictions and short distance read range requirement.

5.2. Mobile device (smartphone) subsystem of RFIDSIM system

The RFIDSIM system adopts Google Nexus S as the RFID-enabled smartphone with NFC technology (see Fig. 2). The Google Nexus S runs the Android operating system. All data in the smart phones are transmitted to the server directly through the web via Wi-Fi. Google Chrome was chosen as the web browser in the RFID-enabled Google Nexus S.

Figure 2. displayed the RFID-enabled smartphone with NFC technology and HF RFID tags using in the study.

5.3. Web portal subsystem of RFIDSIM system

The web portal is an information hub in the RFIDSIM system for a jobsite safety management. The web portal enables all participants to log onto a single portal, and immediately obtain information required for planning. The users can access different information and services via a single front-end on the Internet. For example, a project manager can log onto the portal, enter an assigned security password, and access real-time jobsite safety inspection information and result. The web portal of RFIDSIM system is based on the Microsoft Windows Server 2008 operating system with Internet Information Server (IIS) as the web server. The prototype was developed using ASP.NET, which are easily combined with HTML and JavaScript technologies to transform an Internet browser into a user-friendly interface. The web portal provides a solution involving a single, unified database linked to all functional systems with different levels of access to information.

The following section describes the implementation of each module in the RFIDSIM system.

5.4. Inspection module

Safety staff members can enter inspection results directly via a smartphone. Additionally, smartphone display the inspection checklist in the each jobsite checkpoint location. Safety staff members can record inspection information for conditions, inspection result, descriptions of problems and suggestions that have arisen during the progress. The module has the benefit that safety staff members can enter/edit inspection results, and all records can be transferred between the smartphone and portal by real-time synchronization, eliminating the need to enter the same data repeatedly.

5.5. Progress monitor module

This module is designed to enable project manager and safety staff members to monitor the progress of inspections management. Additionally, project manager and safety staff members can access the progress or condition of jobsite safety checkpoints locations. The progress monitor module provides an easily accessed and portable environment where project manager and safety staff members can trace and record all safety-related inspection information regarding the status of inspections of safety checkpoints locations.

6. Polite test

This study is applied in Taiwan construction jobsite for the polite test. The construction building base was located in the Taipei metropolitan area, near MRT stations and public facilities. The construction project include 30 floors steel reinforced concrete structure buildings, underground parking and five floors of public space. There are three main buildings in the project. The project includes corporate headquarters, office buildings, and a business hotel. This study utilizes an RFIDSIM system in the jobsite safety inspection management. Existing approaches for tracking and managing safety inspection adopt manually updated paper-based records. The most of inspection work were paper-based work by manual entry although construction management system was developed for information management.

However, information collected by staff members using such labor-intensive methods is re-work and ineffective in the inspection results entry. Therefore, jobsite safety management division and safety staff members utilized the RFIDSIM system to enhance safety inspection and jobsite safety management in the case study. HF Passive read/write RFID tags were used in the case study. After the critical safety checkpoint locations were selected, each HF RFID tag for the safety checkpoint location was made, and the unique ID of safety check-point was entered into the RFIDSIM system database. After the safety checkpoint location was assigned to be monitored for safety inspection, the safety checkpoint location was scan-ned with a RFID tag to enter the RFIDSIM system. During the setup phase, all the ID of safe-ty checkpoint in the RFID tag had been determined and entered the database for system, and then the RFID tag was attached in the safety checkpoint location (see Figs. 3 and 4). Fi-nally, the tag will be scanned and checked before the inspection work.

Figure 3. Displayed the safety checkpoint location attached UHF RFID tag in the case study. (A)

Figure 4. Displayed the safety checkpoint location attached UHF RFID tag in the case study. (B)

Before the inspection work, the safety staff members can check the inspection list from smartphone, refer the relative information and can make the preparation work without printing any paper document. During the inspection progress, the safety staff member scanned the RFID tag first and to select the inspection result (see Figs. 5 and 6). The system would update inspection information of jobsite safety checkpoints location via browser under wireless circumstance. After the jobsite safety checkpoints locations were inspected, staff members recorded the status and execute the work by procedure. After the operation, safety staff member recorded the result of inspection, edited the description in the smartphone, and provided the updated information to the system (see Fig. 7). Finally, the safety manager and the authorized staff members accessed the updated information from jobsite office synchronously. Fig. 8 displayed the process flowchart of RFIDSIM system. Fig. 9 displayed safety staff used RFID-enabled smartphone to scan RFID tags and edited the description in the smartphone. Fig. 10 displayed project manager entered the RFIDSIM system and accessed the safety inspection result.

Figure 5. Displayed safety staff used RFID-enable smartphone to scan RFID tag. (A)

Figure 6. Displayed safety staff used RFID-enable smartphone to scan RFID tag. (B)

Figure 7. Ssplayed the staff updated the inspection information in jobsite checkpoint location

7. Field tests and results

Overall, the field test results indicate that HF passive RFID tags are effective tools for jobsite safety management in construction. The RFIDSIM system was installed on main server in the jobsite office. During the field trials, verification and validation tests were performed to evaluate the system. The verification aims to evaluate whether the system operates correctly according to the design and specification; and validation evaluates the usefulness of the system. The verification test was carried out by checking whether the RFIDSIM system can perform tasks as specified in the system analysis and design. The validation test was undertaken by asking selected case participants to use the system, and provide feedback by answering a questionnaire. Some comments for future improvements of RFIDSIM system were also obtained from the case participators through user satisfaction survey. Table 1 shows a comparison using a traditional paper-based inspection approach and the proposed system. The next section presents the detailed results of the performance evaluation and the user survey conducted during the field trials.

Item	Traditional Approach Method	Proposed Approach Method
Inspection recording method	Paper forms	Used RFID-enabled smartphone to scan RFID tag
Inspection data search speed	Referring to Inspected item and checklist	Use electronic forms
Inspector location tracking	Paper forms	Real-time Update database
Is inspection on schedule?	Hard to check	The system record automatically
Inspection history management	Paper forms	Access the system and refer directly
Information Dissemination	Paper forms delivering	Access the system directly and share information

Table 1. System Evaluation Result

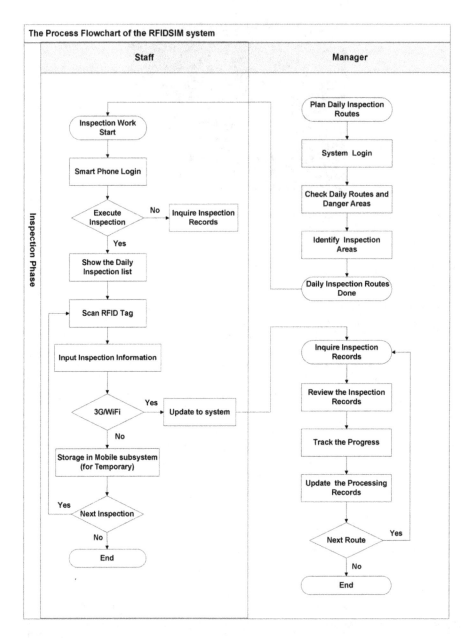

Figure 8. displayed the process flowchart of RFIDSIM system.

Figure 9. displayed safety staff used RFID-enabled smartphone to scan RFID tags and edited the description in the smartphone.

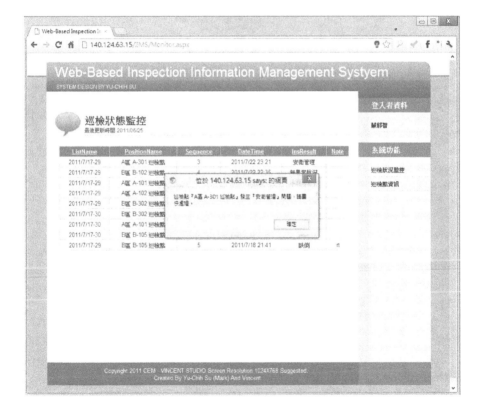

Figure 10. displayed project manager entered the RFIDSIM system and accessed the safety inspection result.

Based on the result obtained from user satisfaction survey indicates that the RFIDSIM system is quite adaptable to the current jobsite safety management practices in construction jobsite, and is attractive to users. This result implies that the RFIDSIM system was well designed, and could enhance the current time-consuming jobsite safety inspection management process. Furthermore, safety staff members just scan the RFID tag and send them electronically to the RFIDSIM system. No additional work was required for any documentation or safety inspection analysis after the data collection.

The advantages and disadvantages of RFIDSIM system identified from the polite test studies application are identified. However, safety staff members' satisfaction survey agree that the RFIDSIM system is useful for improving the efficiency and effectiveness of automated data acquisition and information collection in important check locations, thus assisting safety managers and safety staff members in managing and monitoring the safety inspection processes of the building. Thus, HF passive tags are suited to jobsite safety management.

The use of RFID and web technology to collect and capture information significantly enhanced the efficiency of jobsite safety inspection processes in construction jobsite. Mobile RFID-enabled smartphone with NFC technology and tags are widely thought likely to improve in the future and significantly improving the safety inspection processes efficiency.

In the cost analysis, the HF tags adopted in this study cost under $1 US dollars each in 2012. The cost of these tags is decreasing every year. The total cost of the equipment applied in this study was $3250 US dollars (including RFID-enabled smartphone with NFC technology reader and one server personal computer). Even the reader initial cost is higher, but it is function expandable and really decreases human work. Experimental results demonstrate that RFIDSIM system can significantly enhance the safety inspection processes efficiency. The use of RFID significantly decreases the overall safety inspection operation time and human cost.

The findings of this polite test revealed several limitations of the RFIDSIM system. The following are inherent problems recognized during the case study.

- RFID tags attached to outdoor checkpoint are easily damaged because of external environmental pollution (such as dust, rain, etc.). Therefore, it is necessary to consider and enhance protection and waterproof of RFID tags.

- The cost of system implementation is high because of the non-permanent facilities of the site environment. The required cost increases because that most the temporary facilities, inspection checkpoints always change the location.

- The RFIDSIM is required by the WiFi wireless network or 3G communications to transfer data. Sometime web-based communications in the jobsite safety checkpoints locations are weak and cause information disconnection because of the poor environment (such as underground, corner, etc.).

8. Conclusions

This study presents a Mobile RFID-based safety inspection management (RFIDSIM) system that incorporates RFID technology and mobile devices to improve the effectiveness and convenience of information flow during construction phase of construction project. The RFID-SIM system not only improves the acquisition of data on safety inspection result efficiency using RFID-enabled smartphones, but also provides a real time service platform during safety inspection progress. In the case study, plugging a RFID scanner into a smartphone creates a powerful portable data collection tool. Additionally, RFID readings increase the accuracy and speed of information search, indirectly enhancing performance and productivity. Safety staff members use RFID-enabled smartphones to enhance seamlessly inspection work processes at checkpoints locations, owing to its searching speed and ability to support detail information during the process. Meanwhile, on the server side, the RFIDSIM system offers a hub center to provide jobsite management division with real-time to monitor the jobsite safety progress. In a case study, the application of the RFIDSIM system helps to improve the process of jobsite safety inspection and management work for the construction jobsite in Taiwan. Based on experimental result, this study demonstrated that HF passive RFID technology has significant potential to enhance jobsite safety inspection and management work in construction management. The integration of real-time inspection information from jobsite safety checkpoints helps safety staff members to track and control the whole inspection management progress. Compared with current methods, the combined results demonstrate that, an RFIDSIM system can be a useful RFID-based jobsite safety inspection management platform by utilizing the RFID approach and web technology.

Building information modeling (BIM) is one of the most promising recent developments in the AEC industry. In the future, application of BIM can be considered and integrated for better and advanced jobsite safety inspection and management. Furthermore, the application of BIM will be a viable approach to jobsite safety management during the construction phase of a construction project. The BIM approach, which is utilized to retain visual status of safety condition in a digital format, facilitates effective safety management in the 3D CAD environment. The BIM provides users with an overview of current jobsite safety inspection result during a given construction project, such that users can track and manage jobsite safety inspection result virtually.

9. Recommendations

Recommendations for implementing the proposed system in the future are given below.

- Cost is a currently significant factor limiting the widespread use of RFID tags in the construction industry. Passive tags are cheaper than active tags. Therefore, passive tags are suited to jobsite safety management.

- The smartphone screen is not large enough for operating the RFIDSIM system fluently. The system should be redesigned and developed to be suitable for the smartphone screen.

- In this study, the major characteristic is that users can apply the RFIDSIM system without purchasing additional RFID reader. Currently, this system is developed for Google Android system and will be developed for Apple i-phone system in the future.

Author details

Yu-Cheng Lin, Yu-Chih Su, Nan-Hai Lo, Weng-Fong Cheung and Yen-Pei Chen

National Taipei University of Technology, Civil Engineering, Taiwan

References

[1] Baldwin, A. N., Thorpe, A. and Alkaabi, J. A. (1994), "Improved material management through bar-code: results and implications of a feasibility study," Proceedings of the institution of Civil Engineers, Civil Engineering, 102(6), 156-162.

[2] Chin, S., Yoon, S., Choi, C., and Cho, C. (2008). "RFID+4D CAD for progress management of structural steel works in high-rise buildings,"Journal of Computing in Civil Engineering, ASCE, 22(2), 74-89.

[3] Elghamrawy, T. and Boukamp, F. (2010). "Managing construction information using RFID-based semantic contexts," International Journal of Automation in Construction, 19(8), 1056-1066.

[4] Elzarka, H. M. and Bell, L. C. (1997), "Development of Pen-Based Computer Field Application," Journal of Computing in Civil Engineering, ASCE, 11(2), 140-143.

[5] Ergen, E., Akinci, B., and Sacks, R. (2006) "Tracking and locating components in a precast storage yard utilizing radio frequency identification technology and GPS," International Journal of Automation in Construction, doi:10.1016/j.autcon. 2006.07.004.

[6] Fayek, A., AbouRizk, S. and Boyd, B. (1998), "Implementation of automated site data collection with a medium-size contractor," in Proc. ASCE Computing in Civil Engineering, Boston, MA, 454-6.

[7] Goodrum, P. M., McLaren, M. A., and Durfee, A. (2006) "The application of active radio frequency identification technology for tool tracking on construction job sites," International Journal of Automation in Construction, 15(3), 292-302.

[8] Jaselskis, E. J. and Anderson, M. R. (1995). "Radio-Frequency Identification Applications in Construction Industry," Journal of Construction Engineering and Management, 121(2), 189-196.

[9] Jaselskis, E. J. and El-Misalami, Tarek (2003). "Implementing Radio Frequency Identification in the Construction Process," Journal of Construction Engineering and Management, 129(6), 680-688.

[10] Lahiri, Sandip (2005), RFID Sourcebook, Prentice Hall PTR.

[11] Lee, Ung-Kyun, Kang, Kyung-In, and Kim, Gwang-Hee (2006). "Improving Tower Crane Productivity Using Wireless Technology." Journal of Computer-Aided Civil and Infrastructure Engineering, Vol. 21, pp.594-604.

[12] Manish Bhuptani and Shahram Moradpour (2005), RFID Field Guide : Deploying Radio Frequency Identification Systems, Prentice Hall PTR.

[13] McCullouch, B. G. (1997), "Automating field data collection in construction organizations," in Proc. ASCE Construction Congress V, Minneapolis, MN, 957-63

[14] Pena-Mora, F. and Dwivedi, G. D. (2002), "Multiple Device Collaborative and Real Time Analysis System for Project Management in Civil Engineering," Journal of Computing in Civil Engineering, ASCE, 16(1), 23-38.

[15] Song, J., Haas, C. T. and Caldas, C. (2006). "Tracking the Location of Materials on Construction Job Sites," Journal of Construction Engineering and Management, 132(9), 680-688.

[16] Song, J., Haas, C. T., Caldas, C., Ergen, Esin, and Akinci, B. (2006). "Automating the task of tracking the delivery and receipt of fabricated pipe spools in industrial projects," International Journal of Automation in Construction, 15(2), 166-177.

[17] Sunkpho, Jirapon and Garrett, J. H., Jr. (2003), "Java Inspection Framework: Developing Field Inspection Support System for Civil Systems Inspection," Journal of Computing in Civil Engineering, ASCE, 17(4), 209-218.

[18] Tserng, H. P., Dzeng, R. J., Lin, Y. C. and Lin, S. T. (2005). "Mobile Construction Supply Chain Management Using PDA and Bar Codes." Journal of Computer-Aided Civil and Infrastructure Engineering, Vol. 20, pp.242-264.

[19] Wang, L. C., Lin, Y. C. and Lin, P. H. (2006). "Dynamic Mobile RFID-based Supply Chain Control and Management System in Construction." International Journal of Advanced Engineering Informatics - Special Issue on RFID Applications in Engineering, Vol. 21 (4), pp.377-390.

[20] Ward, M. J., Thorpe, A. and Price, A. D. F. (2003), "SHERPA: mobile wireless data capture for piling works, "Computer-Aided Civil and Infrastructure Engineering, 18, 299-314.

[21] Yagi, Junichi, Arai, Eiji and Arai, Tatsuo (2005). "Construction automation based on parts and packets unification," International Journal of Automation in Construction, 12(1), 477-490.

[22] Yin, Y.L., Tserng, H. P., Wang, J.C. and Tsai, S. C. (2009). "Developing a precast production management system using RFIF Technology," International Journal of Automation in Construction, 18(5), 677-691.

Chapter 5

Interacting with Objects in Games Through RFID Technology

Elena de la Guía, María D. Lozano and
Víctor M.R. Penichet

Additional information is available at the end of the chapter

1. Introduction

Interactive games aimed at educational environments are becoming increasingly important in children's learning. At the same time, technological advances are definitely causing the arrival of new computational paradigms, such as Ubiquitous Computing or Internet of Things. Ubiquitous Computing was defined by Mark Weiser in 1988, which provides the user with advanced and implicit computing, capable of carrying out a set of services of which the user is not aware. Internet of Things is similar to the Ubiquitous Computing paradigm and was introduced by Kevin Ashton in 1999 [7]. The scenario is described as a daily life object network where all of the objects are digitalized and interconnected.

The main objective of this chapter is focused on how to exploit the evolution of technology to improve user interaction in game environments through digitalized objects with identification technology (such as RFID or Near Field Communication). Digitalized objects are used as interaction resources. They are used in conjunction with mobile devices providing the performance of tasks with a simple and intuitive gesture. In the first place, mobile devices offer sophisticated methods to provide users with services to make use of information and to interact with objects in the real world. In the second place, physical objects are associated with digital information through identification technologies such as RFID. In this context, physical mobile interactions allow users to play games through natural interaction with objects in the real world. This chapter has six sections. Section 2 describes some concepts such as: Ubiquitous Computing, the Internet of Things and the types of interaction used in games. Section 3 presents the general infrastructure of RFID systems. In section 4, we describe the development of two RFID games. In section 5 their advantages and disadvantages are presented. Finally, conclusions are set out in Section 6.

2. Related works

Ubiquitous computing involves computers and technology that blend seamlessly into day to day living. Weiser described the concept in the article [8] in 1991.

The idea of a disappearing technology can clearly be applied to the trend in RFID technology development. In recent years, RFID technology was used in retail [2] and logistics [3]. Nowadays RFID Technology is becoming such an ubiquitous technology, it has led to a particular interest in developing a system in smart spaces. The Internet of Things is similar to the Ubiquitous Computing paradigm, which was described by Kevin Ashton in 1999 [7]. This concept refers to the interconnection of everyday objects in a network. i.e., each object such as a table, a chair or a refrigerator may include integrated identification technology. In this way, the Internet evolves from traditional devices to real objects thanks to the use of technologies such as wireless sensors or RFID.

In this chapter we have focused on games as an educational tool for children's learning. A video game is a software programme created for entertainment and learning purposes in general. It is based on the interaction between one or more people and an electronic device that executes the game. Over the past decades, video games have become a mainstream form of entertainment and communication which are highly accepted and successful in the society. People like playing games for several reasons: as a pastime, as a personal challenge, to build skills, to interact with others, for fun, or as tool for learning. In recent years, the advancement of technology has allowed designs to implement intuitive and new forms of interaction between the user and the console. Some of the devices used are: Kinect, Wii, Multi-Touch Technology, Virtual Reality, and Identification technologies such as RFID, NFC. The following describes in detail the devices and ways of interacting that there are between systems and users.

Kinect is a motion sensing input device that is connected to the console and PC video. It allows the user to interact with the game through movement and voice. In order to function, it requires technologies such as sensors, multi-array microphone, RGB camera and an internal processor. Some existing games that incorporate this technology with learning games are: [4][5][6]. These games offer a new and attractive interaction technique based on movement and voice. However, the new interaction needs some getting used to, most especially for children who have either physical or cognitive disabilities, as it can be exhausting to play through movement. Another obstacle is the space requirement and the hardware, such as the camera, is more delicate and expensive. Another device developed to improve the interaction between user and console is the *Wii Remote*, which is used as a handheld pointing device and detects movement in three dimensions. This device incorporates technologies such as: accelerometers, Bluetooth...[21].The main problem is the need for battery.

In addition, there is *Virtual Reality* software using helmets, gloves and other simulators. In this way the user may feel more immersed in the game, and it is very engaging and motivating, but the problem is the high cost of devices, and the difficulty in the use of certain devices. Also, an additional person is required to control the players and devices [9][10]. *Multi-*

touch technology for games allows the users to play on digital tabletops that provide both an embedded display and a computer to drive player interactions. Several people can thus sit around the table and play digital games together. This technology uses infrared LEDs and photodiodes, which are discretely mounted around the perimeter of the LCD. The principle of an infrared touch screen is the combination of an infrared (IR) LED and an IR-sensitive photodiode. As soon as there is an object or finger between the LED and the photodiode, the latter no longer detects the IR light from the LED. This information is the basis for the input detection. You can interact with them through multiple objects (including fingers). Some of the games implemented with touch technology for learning are:[11][12][13][14].

Identification technology such as *RFID* and *NFC* has been used to transmit the identity of an object using radio waves. In this way different types of interaction are allowed, such as touching which involves touching an object to a mobile device and enabling the user to perform the selected task. For example [15][16] show some projects using this technique.

- Scanning: the mobile device or other device is capable of scanning information and interacting with the system to provide a service to the user.

- Approach&remove: [17] this is a style of interaction which allows us to control user interfaces of a distributed nature by making a gesture with the mobile device. Interaction, as mentioned previously, may be absent or may simply consist of approaching the mobile device to digitized objects.

In this chapter we propose another kind of interaction, in which the mobile devices are stationary and the user used physical objects for interacting with the display.

Some systems that use identification technology are described as following: Smart Playing Cards [26] is a game based on RFID; this technology is integrated in cards. Augmented toys are digitalized with RFID technology simulating the real world [18] [19]. Meta-Criket is a kit developed for augmenting objects [25]. Hengeveld described in [20] the value of designing intelligent interactive games and learning environments for young children with multiple disabilities to increase their language and communication skills. In [21] we can find a proposal that digitalizes toys to help deaf children to learn sign language. This system [24] focuses on assessment and training for special children, allowing the user to store data through RFID cards data for processing daily and providing treatment advice. However, this project only focuses on monitoring the child and does not take into account activities to improve their intellectual ability. [22] describes a RFID musical table for children or people with disabilities. The table is designed for people who cannot navigate through menus or by using buttons on an iPod, and serves to enable them to select albums or songs from a music list from an iPod Touch. This system is very specific; it is more focused on entertainment. Logan Proxtalker [23] is a communication device which allows any user to communicate with symbols "PECS" System (Picture Exchange Communication), which is a device to retrieve vocabulary stored in different labels in order to play actual words. These systems provide entertainment and user interaction with the environment. The disadvantage is that are very specific and none of them has focused on the stimulation of the cognitive abilities of people with intellectual disabilities.

The advantages offered by these devices and systems are numerous. They enhance positive attitudes in users. They feel more motivated and encouraged to learn. However, the systems present the following disadvantages:

- The user needs a minimum knowledge of computer use. Not everybody can use a computer and some devices, like a mouse or a keyboard are not intuitive for people with cognitive disabilities. They need someone to help them.

- The system requires highly specialized hardware / software which can be expensive (simulators, virtual reality). In some games, impaired users may have difficulties finding specific information.

On the other hand, RFID technology has many benefits over other identification technologies because it does not require line-of-sight alignment, tags can be identified simultaneously, and the tags do not destroy the integrity or aesthetics of the original object. Due to the low cost of passive RFID tags and the fact that they operate without a battery, this technology is ideal for converting a real object in a physical interface capable of interacting with other devices

3. RFID-games proposal

The main objective of the project was to develop educational games for children that offer easy interaction based on RFID. For this purpose, the advantages offered by games developed in the pre-computer age (traditional games) were combined with the advantages and benefits of computer games.

To begin with, there are many advantages of traditional games. These were designed and carried out in the physical world with the use of real-world properties such as physical objects, our sense of space, and spatial relations.

Pre-computer games interactions consisted of two elements: user-to-physical-world interaction and user-to-user interaction. The physical objects were easily assimilated by the children, allowing users to interact intuitively with them.

There are also many benefits of computer games. These are more popular than traditional games. Some the advantages are the following:

- People create the illusion of being immersed in an imaginative virtual world with computer graphics and sound.

- Computer games are typically more interactive than traditional games, which enables the user to feel more motivated.

- Computer games allow feedback to be easily shown, as well as notifications about the game process and other important information.

Taking advantage of real physical objects and the benefits that new technologies offer us, we have designed a new way to interact with the system. It is based on physical objects that integrate RFID technology and allow us to interact with Graphics User Interfaces.

This kind of the system functions as follows: in the main game an interface is projected on the wall. Users with physical interfaces, i.e., the objects that integrate RFID tags, can interact with the main interface; this requires the mobile device that incorporates the RFID reader to interact with the main interface, which is necessary to bring objects to the mobile device (See Figure 1).

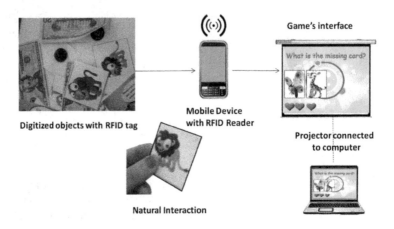

Figure 1. Digitized objects with RFID tags that communicate with the game's interface through the mobile device.

Due to the need to make a simple, accessible and intuitive system and considering the multiple technologies used to develop it, it was decided to follow an architecture based on three layers. The system infrastructure is divided in the following layers: Application Layer, Network Layer and Perception Layer. In the next section, we explain the latter in more detail (See Figure 2).

3.1. Presentation layer

This layer is the intermediary between the user and the system. Its main function is to allow the user to easily interact with the system. In our case study, the games are designed for children and users with special needs and for this reason we must focus primarily on usability and accessibility of the system. The main requirements that have been followed for the development of this type of games are:

Designing simple interfaces so that users do not have to learn to use it, acquire new skills, or need help.

Avoiding distraction and facilitating the interaction so that the user need not know and memorize how the system works.

Avoiding fear of interacting with the system, as well as providing notification of game development and the collaboration of information among players.

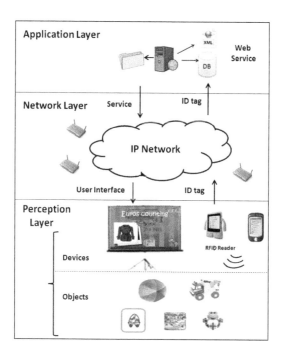

Figure 2. System architecture divided into three layers: Perception, Network and Application

This layer is divided into two parts. Firstly, there are the objects that integrate RFID technology, also called interaction resources, and secondly, there are interaction devices, through which are offered relevant services.

• Objects. Their main function is to facilitate the human-computer interaction. These resources need to have a RFID reader nearby to perform services. The main reason to use objects that interact with the environment is the following: The user uses human factors such as perception in order to interact with the environment. When an object similar to other objects with similar appearance is seen, the mind of the user automatically associates the object with its function.

• Devices. These computing devices are used as input and output of a system. They are communication channels. They are responsible for obtaining information from users without that them being aware of it. In this particular case, a mobile device has been camouflaged in a toy in such a way that it is more engaging and intuitive to users. The devices available in the system are described as following:

Mobiles devices: These devices internally incorporate the RFID reader, allowing users to communicate with the system through RFID technology.

Projector: This shows the game user interface, the results and feedback. The software is run on a PC or laptop. It returns the information in textual and audio format to facilitate the use of games. It works dynamically and responds to the information sent to web services (Application Layer) through the communication network (Network Layer).

The user communication style with the device is very intuitive, which is why no prior knowledge is necessary (see Figure 3), it is only necessary to move the toy, card or object, depending on the game, closer to the mobile device (hidden in an object). The interaction and the processes that occur below the system are implicitly run by the user.

In this case, the collaborative screen shows the game which is being executed. It may show some objects and to associate that object the user has to interact with it, just by moving the corresponding object closer to the mobile device. From this moment all processes are run implicitly. The collaborative screen displays the pictures, text and sounds, depending on the game executed.

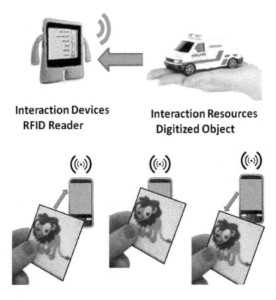

Figure 3. Interaction. The user brings the digitized object closer (interaction resource) to the mobile devices that contain the RFID Reader. This is an interaction device hidden in an object [27].

The communication between interaction devices (mobile devices) and interaction resources (digitalized objects) is the following: The RFID tag (embedded in the object) is a small chip integrated circuit, adapted to a radio frequency antenna that enables communication via radio. The energy to generate communication is received from the reader's radio waves (integrated into the mobile devices).

The device on the client's side includes a reader and a controller that is responsible for processing information received by the physical object and transforming it into useful information, such as an XML message that is sent to the server, which will process the message and trigger an action, such as the generation of user interfaces or the information requested at that time. The network technology is then used to notify the customer with through web services, connecting the two components: the client and the server (See Figure 4).

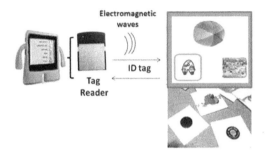

Figure 4. Communication is based on RFID technology. The mobile device has RFID reader inside. It sends electromagnetic waves when a digitalized object is close to mobile device. It processes the information contained in the object and carries out the required action.

3.2. Network layer

This layer enables the information obtained from the perception layer to be transmitted. This layer is composed by different wireless access technologies such as, Wireless Local Area Networks (WLAMs) (IEEE 802.11 variants), Bluetooth (IEEE 802.15.1). Wireless networks are a good option to establish wireless and mobile communications within the Internet of Things. We have used Wi-Fi technology because it allows connection of heterogeneous devices with the system (the computer interface that supports games and mobile devices which communicate with the objects). In addition, it allows user mobility, is highly scalable, efficient and lightweight.

3.3. Application layer

This layer provides services to support the stimulating games. It is consists of a server, which is a computer as part of a network, providing services to the devices which are connected to it. It provides important functions such as Web Services database.

Web Services are a set of protocols and standards used to exchange data between applications in order to offer services. They facilitate interoperability and enable automated services to be offered, automatically causing the generation of user interfaces, thus allowing user consistency and transparency in use of the technology. Web services are of great importance in the trend of distributed computing on the Internet. To broaden and clarify the concept of Web services, we can quote a presentation by Dr. Marcos Escobar: "A Web Service is a software component that communicates with other applications by coding the XML message and sends this message via standard Internet protocols such as HTTP (Hypertext Transfer Protocol)". Intuitively, a Web Service is similar to a Web site that has a user interface that provides a service to applications, by receiving requests through a message formatted in XML (Extensible Markup Language) from an application, it then performs a task and sends a response message, which is also formatted in XML. The standard protocol for messages is SOAP (Simple Object Access Protocol). A SOAP message is similar to a letter: it is an envelope containing a header with the address of the recipient, a set of delivery options (data encryption), and a body with the information or data of the message. The performance of the web services is as follows: the client application sends an XML message to the server, and then the services contained provide an XML document called WSDL (Web Services Description Language). Its aim is to describe in detail the interfaces so that the user can communicate with the service. XML Web services are registered so that the user can easily find them. This is performed using UDDI (Universal Description Discovery and Integration). The response to the customer is another XML message that is capable of generating the user interface that the device in the client's side is going to display at that moment. Figure 3 shows the communication that takes place between Web services and client applications.

Database is an organized collection of data, today typically in digital form. The data is typically organized to model relevant aspects of reality. In this case, the database is composed the idtag field. Each idtag is associated with the web service function. Among the functions are the following: execute a method, update information.

The internal operation is as follows: the web service receives the information, which is the output layer, and specifically the id tag which in this application has been read from mobile devices. The system checks the method associated with this id tag in the database. Web Service receives information about the method that it must execute. The execution of this operation depends on the following parameters: the object identifier, the executed game and the current status in the game. A common flow of actions that a user may perform could include:

• Updating the database and results internally in the system.

• The system automatically generates the corresponding game interface. The projector displays it. According to the action carried out, different messages might be shown.

• If the answer is right, a message indicates the outcome of play. This user interface congratulates and encourages the children to continue playing. A few seconds later, the interface related to the game that is running appears, but at a higher level than before.

- If the answer is wrong, a message indicates the outcome of the play. This user interface motivates and encourages them to try again. The next user interface is related to the game that is running at the time, but at the same level as before. Voices and motivating messages sound in every interface to make the user feel actively accompanied and encouraged.

- The system automatically generates the corresponding mobile user interface. It shows feedback and status of the system according to the action carried out.

4. Case studies developed by using RFID technology

In this section we describe two systems built in the University of Castilla-La Mancha (Albacete). The main objective is to take advantage of RFID technology to build systems that improve the user experience.

We used the same architecture for both games, while changing the contents and taking into account the cognitive abilities that we aimed to stimulate in each particular case.

This system functions by projecting an interface on the wall in the main game. Users with physical interfaces, i.e., the objects that integrate RFID tags, can interact with the main interface; this requires a mobile device that allows the RFID reader to interact with the main interface by bringing an object closer to the mobile device to play the game. For example, if in the game an object must be associated with another, the user only has to bring the corresponding object closer to the mobile device for the system to recognize it and display the outcome of the game.

4.1. Train InAb system

Intellectual disability, also called mental retardation, is a disability characterized by significant limitations in intellectual functioning and in adaptive behavior skills manifested in conceptual, social and practical aspects [1].

So far, this group has always had barriers imposed by society and by technology as it has often not been known how to adapt to the personal needs of each of these people.

Gradually, this situation has been improving with technological assistance and that of society. However, many of these people consider the world of technology to be strange and difficult to use.

TrainInAb (Training Intellectual Abilities) is an interactive and collaborative game designed to stimulate people with intellectual disabilities. The game is based on RFID technology; it allows a new form of human-computer interaction to be integrated. The user can interact with the system through everyday objects such as cards, toys, coins, etc. (See Figure 5). For example, if in the game an object must be associated with another, the user only has to bring the corresponding object closer to the mobile device, which the system will then recognize and display the outcome of the game (See Figure 5 and Figure 6)

The package consists of three different types of game, each aimed at stimulating a different cognitive ability such as memory, calculation, attention and auditory discrimination.

- They are divided into different levels to motivate the child when using the game. If the child fails, s/he loses a life and if the user wins, s/he moves on to the next level. Each level is more difficult.

- It displays the external information differently, as it is different for every level.

- The information is displayed as text, voice and graphics. In addition, the game can show the status and game results when the game ends

- The feedback-state messages are motivating for the user who then feels more encouraged to continue playing.

- The user has the possibility of repeating items.

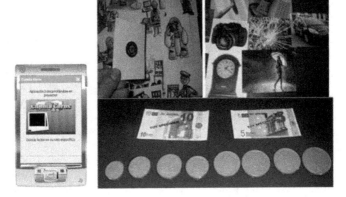

Figure 5. The first image shows the Mobile devices interfaces. The next image shows the Physical user interfaces, that is, objects that integrate the RFID inside.The first objects are cards with images from the game, and the last image shows the notes and coins used for the game.

Figure 6. Main interface of the game designed to stimulate user memory, attention and calculative abilities.

4.2. StiCap

Attention-deficit/hyperactivity disorder (ADHD) is a neurobiological disorder characterized by developmentally inappropriate impulsivity, inattention, and in some cases, hyperactivity. Children who are affected by this disorder have occasional difficulty paying attention or controlling impulsive behavior. This problem affects them in their daily lives at home, at school, at work, and in social settings.

StiCap, Stimulating Capabilities, is an interactive system to improve attention and learning in children with ADHD. It is directed towards psychological therapies, in schools, allowing supervision by professionals, parents, and teachers.

The system consists of three games: two oriented towards memory improvement and another one oriented towards vocabulary enrichment. It is composed of the following devices: cards integrating RFID tags used as interactive resources which allow a one-way transfer of information between a user and the system; mobile devices provide the necessary communication between the cards and the system and a projector or any other big display showing the game interface which is running on any PC or laptop.

Figure 7. Main interface of the game designed to stimulate user memory and attention [28]

5. Benefits and drawbacks

In this section we will discuss the advantages offered by the integration of RFID technology in the new scenarios.

The main advantages of the system are the following:

• Reduction of the cognitive load. This means that users have to rely more on recognition skills than on their memory and that they do not have to remember complicated abbreviations and codes. For this reason, it has been designed in a very graphic way and has also used common objects which can be easily assimilated.

• Flexibility. This refers to the multiple ways in which the user and the system can exchange information. The information exchanged is displayed as text, voice, cheerful sounds or by using graphics. The goal is to adapt to any user, regardless of any disability or limitation he/she may have.

- Flexibility in the number of users. This is a multi-player game. This allows users to share and exchange experiences with other users. The situation of each user may be complex and variable and for this reason, the game can also be used by one player.

- Flexibility in terms of space. Players can be situated anywhere in the room, the only requirement is that the mobile device is connected to the server.

- Very cheap to develop. Mobile devices will incorporate RFID technology in the short term and passive RFID tags are very inexpensive. In our case, only one mobile device is required, which is why it is low cost.

- Expandable. It offers the possibility to extend the games. The topic can be changed easily. The only requirement is that the RFID must be integrated in the object selected.

- Interaction with the system is simple and intuitive. Common items are familiar and can be easily assimilated by users, making it more predictable to use. They do not need prior knowledge of the system or device.

- The cognitive stimulation of the system can enhance mental abilities such as perception, attention, reasoning, abstraction, memory, language, orientation processes, while optimizing their performance. These games can be used as therapy for the cognitive deficit.

Thanks to this technology, the implementation of new interfaces can be developed for any mobile device, allowing system usability and user-friendly interaction, thus improving user satisfaction.

One possible limitation are that it requires connectivity to another network interconnection. The server needs to contain all the data from RFID tags, so in very complex systems we can find a lot of data, which might be difficult to manage.

6. Conclusions

Educational games are currently making a very positive impact and are extremely successful among society, especially among children.

Emerging technologies and mobility are being inserted without society realising by providing services previously unthinkable. In recent years, devices have been invented that offer new techniques for interaction between humans and game consoles. Nowadays, the user can interact through movement, voice command control, virtual reality, mobile devices, etc... However, there are still some hardware limitations for children and especially people who need special education.In recent years, RFID technology is booming and being used to digitalize spaces and objects easily, so we are getting closer to the new paradigm predicted by Weiser, ubiquitous computing.Exploiting the advantages offered by this technology, this chapter proposes a new form of interaction based on objects that integrate RFID technology.

In this way, anyone can interact with the software(in this case with the games)in an intuitive way.

Acknowledgements

This research has been partially supported by the Spanish CDTI research project CENIT-2008-1019, the CICYT TIN2011-27767-C02-01 project and the regional projects with reference PAI06-0093-8836 and PII2C09-0185-1030. I would like to especially thank to Yolanda Cotillas Aranda y Erica González Gutierrez for their collaboration on this project.

Author details

Elena de la Guía, María D. Lozano and Víctor M.R. Penichet

University of Castilla-La Mancha, Spain

References

[1] Luckasson, R., Borthwick-Duffy, S., Buntix, W.H.E., Coulter, D.L., Craig, E.M., Reeve, A., y cols. (2002). Mental Retardation. Definition, classification and systems of supports (10th ed.). Washington, DC: American Association on Mental Retardation.

[2] Wamba, S.F., Lefebvre, L.A., Lefebvre., E.: Enabling intelligent B-to-B eCommerce supply chain management using RFID and the EPC network: A case study int he retail industry. In: The 8th international conference on Electronic commerce.(2006) 281–288

[3] Srivastava, B.: Radio Frequency ID technology: The next revolution in SCM.Business Horizons 47/6 (2004) 60–68

[4] Desiree DePriest, Kaplan University, Florida and Karlyn Barilovits, Walden University, Maryland. LIVE: Xbox Kinect©s Virtual Realities to Learning Games TCC Online Conference, 16th Annual Technology, Colleges an Community Online Conference. Emerging Technologies: Making it Work April 12-14, 2011. Volume 2011, Number 1. (pages 48-54) Full Text (pdf) http://etec.hawaii.edu/proceedings/2011/DePriest.pdf

[5] M. R. Murray, An Exploration of the Kinesthetic Learning Modality and Virtual Reality in a Web Environment, unpublished PhD dissertation, Brigham Young University, Salt Lake City, Utah, 2004.

[6] Lelia Meyer's, Pilot Program Incorporates Video Games into Classroom Learning. The Journal. Transforming Education through Technology http://thejournal.com/arti-

cles/2012/03/13/pilot-program-incorporates-video-games-into-classroom-learn-ing.aspx

[7]   Kevin Ashton: That 'Internet of Things' Thing. In: RFID Journal, 22 July 2009. Re-trieved 8 April 2011.

[8]   Weiser, M. (1991) The Computer for the 21st Century, Sci Amer., ISSN: 1064-4326

[9]   Standen, P.J. and Brown, D.J.. Virtual Reality in the Rehabilitation of People with In-tellectual Disabilities, Review, CyberPsychology & Behavior, vol. 8, no. 3, pp. 272-282, 2005.

[10] Takacs, B. Special Education & Rehabilitation: Teaching and Healing with Interactive Graphics, Special Issue on Computer Graphics in Education IEEE Computer Graph-ics and Applications, pp. 40-48, September/October 2005

[11] Alissa N. Antle, Allen Bevans, Josh Tanenbaum, Katie Seaborn, Sijie Wang, Futura: design for collaborative learning and game play on a multi-touch digital tabletop, Proceedings of the fifth international conference on Tangible, embedded, and em-bodied interaction, January 22-26, 2011, Funchal, Portugal

[12] Alissa N. Antle, Alyssa F. Wise, Kristine Nielsen, Towards Utopia: designing tangi-bles for learning, Proceedings of the 10th International Conference on Interaction De-sign and Children, p.11-20, June 20-23, 2011, Ann Arbor, Michigan

[13] Jacob George, Eric de Araujo, Desiree Dorsey, D. Scott McCrickard, Greg Wilson: Multitouch Tables for Collaborative Object-Based Learning. HCI (9) 2011: 237-246

[14] Bertrand Schneider, Megan Strait, Laurence Muller, Sarah Elfenbein, Orit Shaer, Chia Shen, Phylo-Genie: engaging students in collaborative 'tree-thinking' through tab-letop techniques, Proceedings of the 2012 ACM annual conference on Human Factors in Computing Systems, May 05-10, 2012, Austin, Texas, USA

[15] Broll,G. Graebsch,R., Holleis,P., Wagner,M. Touch to play: mobile gaming with dy-namic, NFC-based physical user interfaces, Proceedings of the 12th international con-ference on Human computer interaction with mobile devices and services, September 07-10, 2010, Lisbon, Portugal

[16] Hardy,R., Rukzio,E. Touch & interact: touch-based interaction of mobile phones with displays, Proceedings of the 10th international conference on Human computer inter-action with mobile devices and services, September 02-05, 2008, Amsterdam, The Netherlands

[17] Romero,S., Tesoriero,R., González,P., Gallud,J. A., Penichet, V. M. R.: Sistema Inter-activo para la Gestión de Documentos Georeferenciados basado en RFID. Interacción 2009, X Congreso Internacional de Interacción Persona-Ordenador. Barcelona. Sep-tiembre 2009. ISBN-13:978-84-692-5005-1

[18] S. Hinske, M. Langheinrich, and Y. Alter, "Building rfid-based augmented dice with-perfect recognition rates," in Proceedings of Fun and Games 2008, Eindhoven, The-Netherlands, LNCS, (Berlin Heidelberg New York), Springer, Oct. 2008.

[19] S. Hinske and M. Langheinrich, "Using a movable rfid antenna to automatically de-termine the position and orientation of objects on a tabletop," in Proceedings of Eu-roSSC 2008, Zurich, Switzerland, LNCS, (Berlin Heidelberg New York), Springer,Oct. 2008.

[20] Lewis, C.: Hci and cognitive disabilities. Interactions 13(3), 14–15 (2006)

[21] Parton et al., Parton BS, Hancock R, duBusdeValempr AD. Tangible manipulatives and digital content: the transparent link that benefits Young deaf children. In: Pro-ceedings of the International conference on interaction design and children. 2010

[22] Music Table.http://www.engadget.com/2011/10/05/arduino-ipod-and-rfid-make-be-atiful-handicapped-accesible-musi/+

[23] Logan Proxtalker. http://www.proxtalker.com/

[24] K. J. Yeo, E. Supriyanto, H. Satria, M. K. Tan, E. H. Yap, Interactive Cognitive Assess-ment and Training support system for special children. Proceedings of the 9th WSEAS International Conference on Telecommunications and Informatics, April 2010

[25] F. Martin, B. Mikhak, and B. Silverman, "Metacricket: A designer's kit for making computational devices," IBM Systems Journal, vol. 39, no. 3&4, p. 795, 2000.

[26] K. R°omer and S. Domnitcheva, "Smart playing cards: A ubiquitous computing game,"Personal and Ubiquitous Computing, vol. 6, no. 5/6, pp. 371–377, 2002.

[27] Pearson, E. and Bailey, C. (2007). Evaluating the potential of the. Nintendo Wii to support disabled students in education. ASCILITE. 2007, Singapore. 19

[28] Guía, E.d.l. Lozano, M. D., Penichet, V.M.R.,"New Ways of Interaction for People with Cognitive Disabilities to Improve their Capabilities",Proc. of 2nd Workshop on Distributed User Interfaces in conjunction with CHI 2012, May 05, 2012, Austin, Texas, USA. I.S.B.N.: 84-695-3318-5

[29] Guía, E.d.l. Lozano, M. D., Penichet, V.M.R.," Stimulating Capabilities: A Proposal for Learning and Stimulation in Children with ADHD", Interaction Design in Educa-tional Environments (IDEE 2012), ICEIS June 2012 Wrocław (Poland)

Chapter 6

Manufacturing Logistics and Packaging Management Using RFID

Alberto Regattieri and Giulia Santarelli

Additional information is available at the end of the chapter

1. Introduction

The chapter is centred on the analysis of internal flow traceability of goods (products and/or packaging) along the supply chain by an Indoor Positioning System (IPS) based on Radio Frequency IDentification (RFID) technology.

A typical supply chain is an end-to-end process with the main purpose of production, transportation, and distribution of products. It is relative to the products' movements from the supplier to the manufacturer, distributor, retailer and finally to the end consumer. Moreover, a supply chain is a complex amalgam of parties that require coordination, collaboration, and information exchange among them to increase productivity and efficiency [1, 2]. A supply chain is made up of people, activities, and resources involved in moving products from suppliers to customers and information from customers to suppliers. For this reason, the traceability of logistics flows (physical and information) is a very important issue for the definition and design of manufacturing processes, improvement of layout and increase of security in work areas.

European Parliament (Regulation (EC) No. 178/2002) [3] makes it compulsory to trace goods and record all steps, used materials, manufacturing processes, etc. during the entire life cycle of a product [4]. According to the European Parliament, companies recognize the need and importance of tracing materials in indoor environments.

Traditionally, the traceability system is performed through the asynchronous fulfilment of checkpoints (i.e. doorways) by materials. In such cases, the tracking is manual, executed by operators. Often companies are not aware of the inefficiencies due to these systems of traceability such as low precision and accuracy in measurements (i.e. no information between doorways), more time spent by operators and costs (due to the full-effort of operators who have to trace target positions and movements). According to [5] every day millions of transport units (cases, boxes, pallets, and containers) are managed worldwide with limited or even with lack of knowledge regarding their status in real-time. In order to overcome the lack of data due to traceability, automatic identification procedures (Auto-ID) could be a solution. They have become very popular in many service industries, purchasing and distribution logistics, manufacturing companies and material flow systems. Automatic identification procedures provide information about people, vehicles, goods, and products in transit within the company [6]. It is possible to note several advantages using an automatic identification system such as the reduction of theft, increase of security during the transport and distribution of assets, and increase of knowledge of objects' position in real-time.

Automatic identification procedures can also be applied to packaging products, instead of to each item contained in the package. Packaging is becoming the cornerstone of processing activities [7]. Sometimes products are very expensive and packages contain important and critical goods (for example dangerous or explosive materials) and the tracking of goods – and packaging in particular – is a critical function. The main advantage of automatic system application to packages is the possibility to map the path of all items contained into the packages and to find out their real-time position. The installation of automatic systems in packages allows costs and time to be reduced (by installing, for example, the tag directly on the package instead of on each product contained inside the package).

The purpose of the chapter is to provide an innovative automatic solution for the traceability of *everything that moves* within a company, in order to simplify and improve the process of logistics flow traceability and logistics optimization. The chapter deals with experimental research that consists of several tests, static and dynamic, tracing the position (static) and movements (dynamic) of targets (e.g. people, vehicles, objects) in indoor environments. In order to identify the best system to use in the real-time traceability of products, the authors have chosen Real Time Location Systems (RTLSs) and, in particular, the Indoor Positioning Systems (IPSs) based on Radio Frequency IDentification (RFID) technology. The authors discuss the RFID based system using UWB technology, both in terms of design of the system and real applications.

The chapter is organized as follows: Section 2 briefly describes IPS systems, looking in more depth at RFID technology. After that the experimental research with the relative results and discussion are described in Section 3. Section 4 presents an analysis of RFID traceability systems applied to packaging. Conclusions and further research are discussed in Section 5.

2. Background of Indoor Positioning System (IPS)

This section presents a general description of IPSs. First, the authors describe logistics flows (physical and informative). After that, the section moves on to describing IPSs (methods for determining the position of a target, criteria to evaluate IPSs, classification of IPSs), underlining the advantages of using automatic identification procedures for tracing objects. Finally, the section provides a brief description of RFID and in particular RFID-UWB technology (Radio Frequency IDentification-Ultra Wide Band).

2.1. Logistics flows

Generally, companies provide goods and/or services to customers, purchasing raw materials from suppliers. In order to increase productivity and efficiency within the supply chain, the parties (suppliers, manufacturers, and customers) have to exchange materials and information among themselves.

In a typical supply chain, logistics flows can be classified into *physical* and *informative*. Physical flows include operative activities (e.g. transport, storage of raw materials, semi-finished and finished products, etc.). A great purpose of the optimization of these flows is the reduction of transport and storage costs. Information flows concern the information on the demand, logistics, and production planning. Figure 1 shows a graphical representation of a supply chain, underlining physical and informative flows.

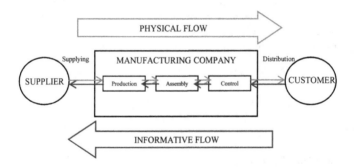

Figure 1. General scheme of a supply chain, underlining materials and information flows

Within the supply chain, it may be essential to know both the position and the movements of operators, pallets, tools, and packages. The traceability of flows within a company is a crucial aspect that has to be optimized.

Traditionally, the process of traceability of goods is performed through the asynchronous and automatic fulfilment of doorways by materials (e.g. bar code reading process) or totally manual by an operator who identifies and measures all movements between work centres, assembly and control workstations, and warehouses (Spaghetti Chart and From-To Chart are two technologies in which the presence of an operator to identify the position and map the movements of goods is necessary). This system implies approximate measurements, full-time effort and wasted time by the operator, and the possibility of human errors. In order to improve performances in the traceability process and to reduce costs optimizing the internal flows, companies are beginning to use automatic identification procedures (Auto-ID). The main advantage of this method is the time reduction in measuring the position and mapping the movements of an object. Real Time Locating System (RTLS) is an automatic system for identifying the real-time position of objects and IPS is the RTLS technology chosen by the authors for developing the experimental research.

2.2. Indoor Positioning System (IPS)

In recent years, indoor location sensing systems have become very popular [8] for locating the position and mapping the movements of goods and people. An Indoor Positioning System (IPS) is a process that continuously determines in real-time the position of something or someone in a physical space (e.g. the location detection of products stored in a warehouse, medical equipment in a hospital, luggage in an airport) [9]. According to [10], an IPS can provide different kinds of data for location-based applications. Any positioning system has at its core the measurement of a number of observable parameters (e.g. angles, velocity, ranges, and range differences) [11]. From the definition by Hightower [9], an IPS works all the time unless the user turns off the system, offers updated position information on the object, estimates position within a maximum time delay, and covers the expected area in which users need to use IPS [10].

In general, a real-time location system is a combination of hardware and software, continuously used to determine and provide the real-time position of assets and resources equipped with devices designed to operate with the system. A location may be described through relative position data with indication of distances, or absolute position data, with some accuracy in any defined grid of coordinates. Generally, location and ranging are reported visually, mostly referring to a map of land, a plan of a building, or in a graph. Alternatively, a change of location may be indicated with sound signals. In particular, a real-time location system uses sensors to determine the real-time coordinates of a tag, everywhere within the area of interest [11]. Curran et al. [11] describe the main industrial applications of indoor location determination systems for companies, in particular the real-time identification of the position of materials, the path control of material flows and warehousing.

Another important industrial application of location positioning system is the *traceability of packages*. Many companies need to track packages, first without the product and after with the products inside, to know the real path (and cost) of their material flows, allowing control of the Work in Progress (WIP) and finally to reduce costs of the system.

2.2.1. Positioning algorithms using IPSs

According to [11], there are several methods for locating and determining the position and movements of an object. A positioning location system can use only one method or combine a number of techniques to achieve better performance. The most commonly used methods are [8]:

- *Triangulation*: this uses the geometric properties of triangles to estimate the target location. It has two derivations: *lateration* and *angulation*.

  - *Lateration* estimates the position of an object by measuring its distance from multiple reference points (it is also called the range measurement technique). According to [8] this method implies the measurement of the *Time Of Arrival* (TOA, that is the travel time of the distance that divides the receiver and the transmitter, knowing the speed of signal propagation) or the *Time Difference Of the signal's Arrival* (TDOA, that is the distance of the difference between the arrival time of signals sent by the transmitter). The distance is derived by computing the attenuation of the emitted signal strength or by multiplying the radio signal velocity and the travel time;

   o *Angulation* (called also *Angle of Arrival* – AOA) is a method that locates the object to be measured through the intersection of several pairs of angle direction lines, each formed by the circular radius from a base station to the mobile target [8]. The main advantages are that a position estimate may be determined with as few as three measuring units for 2D positioning, and that no time synchronization between measuring units is required. The disadvantages include relatively large and complex hardware requirements and location estimate degradation as the mobile target moves away from the measuring units [8];

• *Scene analysis:* this refers to the type of algorithms that first collect the features (*fingerprints*) of a scene and then estimate the location of an object by matching online measurements with the closest *a priori* location fingerprints [8]. Location fingerprints refer to techniques that match the fingerprint of some characteristics of a signal that is location dependent. The location fingerprint is based on two moments: the offline phase, in which an analysis of the measuring environment is conducted, collecting a large number of coordinates, and the online phase, in which target data is compared with that collected before and the location is identified with the point with the most similar values [8]. This technique is subjected to signal interferences, because of obstacles presented in the environment;

• *Proximity* is the simplest method of positioning, but it can only provide an approximate location of the target, and not an absolute position. Proximity algorithms provide symbolic relative location information. Usually, this relies on a dense grid of antennas, each having a well-known position. When a mobile target is detected by a single antenna, it is considered to be located with it. When more than one antenna detects the mobile target, it is considered to be located at the one that receives the strongest signal. This technique can be implemented over different types of physical media. In particular, systems using RFID are often based on this method [8].

2.2.2. Evaluation criteria for IPS systems

In order to evaluate the performance of an IPS, various system performance and deployment criteria are proposed:

• *Accuracy* (or location error) is the most important requirement of a positioning system [8]. Usually, mean distance error is adopted as the performance metric, which is the average Euclidean distance between the estimated and true location. The higher the accuracy, the better the system. Accuracy alone, however, is not sufficient to completely define the performance of a positioning system and, as such, a trade-off between "suitable" accuracy and other characteristics is needed;

- *Precision* is the success probability of position estimation with respect to predefined accuracy [10] and considers how consistently the system works. Precision is a measure of the robustness of the positioning technique as it reveals the variation in its performance over many trials. In order to measure the precision of a system, the cumulative probability functions of the distance error is used;

- *Complexity* of a positioning system can be attributed to hardware, software, and operational factors. In particular, the software complexity is the computing complexity of positioning algorithms. Elements that influence the complexity are human efforts during the initialization and maintenance phases, and the computing time requested of the tag by the operator to determine the target position [8];

- *Robustness* is the ability of an IPS to keep operating even in serious cases, such as when some devices in the system are malfunctioning or damaged, or some mobile devices run out of battery power [10];

- *Scalability* is the ability to function normally when the positioning scope is large. Usually, the positioning performance degrades when the distance between the transmitter and the receiver increases. A location system may need to scale on two axes: geography (the covered area or volume) and density (the number of units located per unit geographic area/ space per time) [8];

- *Cost* of a positioning system may depend on many factors, such as money, time, space, weight, and energy. The time factor relates to installation and maintenance. The space factor is linked to the space and weight constraints of system units. Energy is an important cost factor of a system: some mobile units are completely energy passive and only respond to external fields, therefore could have an unlimited lifetime. Other mobile units have a lifetime of several hours after which they have to be recharged or the battery needs replacing [8].

2.2.3. IPSs classification

According to [10], there are several criteria for classifying an IPS. One criterion is based on whether an IPS uses an existing wireless network infrastructure to measure the position of an object. IPSs can be grouped as *network-based* and *non-network-based* approaches. The network-based approach takes advantages of the existing network infrastructure, where no additional hardware infrastructures are needed. For cost reasons this approach is preferred. However, the non-network-based approach uses dedicated infrastructures for positioning and has freedom of physical specifications by the designers, which may offer higher accuracy.

More generally, IPSs are classified according to the method used to determine the target position. Figure 2 ([12] version modified by [8]) shows the technologies used to determine the target position according to *resolution* (the performance of IPSs) and *scalability* (the environment in which each technology is best suited).

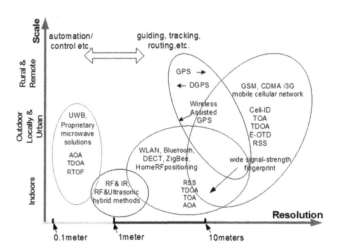

Figure 2. IPSs classification by *resolution* and *scalability* ([12] version modified by [8])

According to resolution and scalability, IPSs can be classified into several groups of automatic positioning systems. The most important are as follows:

- *Infra-Red (IR) based systems* are the most common positioning systems, since IR technology is available on board various wired and wireless devices, such as TVs, printers, mobile phones, etc. [13, 14]. They have several advantages such as wide availability, great positioning accuracy, simple system architecture and light and small tags. In addition, since the whole infrastructure is very simple, it does not need costly installation and maintenance [15]. The line-of-sight requirement and short-range signal transmission are two major limitations that suggest it is less effective in practice for indoor locations [16]. IR systems require the absence of interference and obstacles between the target and the sensor. For these reasons, they cannot be applied to some kinds of indoor scenarios in which the environment is complex;

- *Ultra-sound positioning systems* use diffusion, refraction, and diffraction phenomena, defined by the parameters of frequency, wavelength, speed of propagation and attenuation. Ultra-sound positioning systems are cheap solutions and their accuracy is high, but their precision is low when compared to IR-based systems, because of the reflection influence [15];

- *Radio Frequency (RF) based systems* are technologies used in IPSs, that can uniquely identify people or objects tracked in the system. They provide some advantages as follows. Radio waves can easily travel through walls and human bodies, thus the positioning system has a larger coverage area and needs less hardware compared to other systems. RF-based positioning systems can reuse existing RF technology systems [17]. They can cover large distances, since they use electromagnetic transmissions and are able to penetrate opaque objects such as people and walls. WLAN (Wireless Local Area Network), Bluetooth, Wireless sensor networks and RFID-UWB (Radio Frequency IDentification-Ultra Wide Band) are based on this technology [15], briefly described below.

o *WLAN* technology is very popular and has been implemented in public areas such as hospitals, train stations and universities. WLAN based positioning systems reuse existing WLAN infrastructures in indoor environments, which lower the cost of indoor positioning. The accuracy of location estimations based on the signal strength of WLAN signals is affected by various elements in indoor environments such as the movement and orientation of human bodies, nearby tracked mobile devices, walls, doors. RADAR system, Ekahau positioning system and COMPAS are the main techniques based on the WLAN positioning technology [18];

o *Bluetooth* is a technical and industrial method for transmitting data in a WPAN (Wireless Personal Area Network). It enables a range of 100 m communication to replace the IR ports mounted on mobile devices [19];

o *Wireless sensor networks* are devices exposed to physical or environmental conditions including sound, pressure, temperature and light, and they generate proportional outputs [20];

o *RFID-UWB* is a method for storing and retrieving data through electromagnetic transmission to an RF compatible integrated circuit [16]. RFID-UWB technology will be explained in detail in the next paragraph.

2.3. Radio Frequency IDentification (RFID)

In recent years, the application of RFID has attracted considerable interest among scientists as well as managers faced with the problem of optimizing production processes in several industries [6]. RFID has enormous economic potential, which many manufacturers (e.g. Wal-Mart, Tesco, Marks & Spencer and other retailers [21-23]) have already recognized and started to use successfully [24].

The main use of RFID systems in industrial applications deals with asynchronous identification. The traditional barcode labels that triggered a revolution in identification systems are inadequate in an increasing number of cases. Barcodes may be extremely cheap, but their limitations are their low storage capacity and the fact they cannot be reprogrammed [6]. A barcode is an optical machine-readable representation of data, which shows data about the object to which it is attached. Unlike an RFID, a barcode represents data by varying the widths and spaces of parallel lines, and may refer to a linear or one-dimension (1D).

Radio frequency identification is a method for storing and retrieving data through electromagnetic transmission to an RF compatible integrated circuit [16]. RFID positioning systems are commonly used in complex indoor environments and their function is to identify an object through radio frequency transmission. The main purpose of this technology is to assume information about animals, objects, or people identified by small tools in radio frequency associated to them. According to [25] some of the more transparent advantages of RFID are as follows:

- RFID does not require line-of-sight to capture data, hence saving time and labour by eliminating the need for unloading a pallet and identifying the load;

- RFID is able to read the contents of an entire pallet load or SKU (Stock Keeping Unit) in seconds and saves time and labour;

- RFID sensors can read data from tags from several meters away;

- Each RFID tag has a unique code;

- RFID can be a read/write system so data can be updated through the supply chain, providing insight into possible trouble spots in distribution, such as theft and damage.

On the other hand, RFID method is an expensive solution, but this limitation could be overcome with better performances of RFID systems.

By describing RFID components and their functions, it is possible to understand the technology and issues that influence the application of an RFID system. A typical RFID consists of three components:

- *RFID tag* (transponder) is the data-carrying device located on the object to be identified. RFID tags are categorized as either passive or active.

  o *Passive RFID* tags operate without a battery. They are mainly used to replace traditional barcode technology and are much lighter, smaller in volume, and less expensive than active tags. They reflect the RF signal transmitted by a reader, and add information by modulating the reflected signal [8];

  o *Active RFID* tags are small transceivers, which can actively transmit data in response to an interrogation. The frequency ranges used are similar to the passive RFID case except for the low and high frequency ranges. The advantages of an active RFID tag are the smaller antenna and the much longer range than passive tags (which can be 10 m). Active tags are ideally suited for the identification of high-unit-value products moving through a harsh assembly process [8];

- *RFID reader* (interrogator) has the overall function of reading and translating data emitted by RFID tags. Readers can be quite sophisticated, all depending on the type of tags that are supported and functions they need to perform. As a result, the capabilities and sizes of readers depend on the application [25]. A reader typically contains a radio frequency module (transmitter and receiver), a control unit, and a coupling element to the transponder. In addition, many readers are fitted with an additional interface to enable them to forward the data received to another system (PC, robot control system, etc.);

- *Host computer* communicates with the reader and information management system.

The RFID components and their connections are shown in Figure 3 ([6] version modified by [25]).

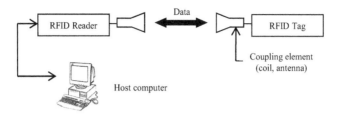

Figure 3. Components of an RFID system ([6] version modified by [25])

2.3.1. RFID – Ultra Wide Band (UWB)

Amongst RFID technologies, Ultra Wide Band (UWB) is the most accurate and fault tolerant system. It can have a widespread usage in indoor localizations.

RFID-UWB is an emerging radio technology marked by accuracy in the estimation of the position, and the precision with which it is possible to obtain that accuracy.

According to the most influential and widespread definition, provided by the *Federal Communications Commission Regulation* [26], an RFID-UWB system is defined as any intentional radiator having a fractional bandwidth greater than 20% or an absolute bandwidth greater than 500 MHz. These requirements mean that a band-limited signal, with lower frequency f_L and upper frequency f_H, must satisfy at least one of the following conditions (Equation 1, 2):

$$\frac{2(f_L - f_H)}{(f_L + f_H)} > 20\% \tag{1}$$

$$f_L - f_H > 500\, MHz \tag{2}$$

According to [27], the main characteristics of an RFID-UWB are the transmission of a signal over multiple frequency bands simultaneously and the brief duration of that transmission. RFID-UWB requires a very low level of power and can be used in close proximity to other RF signals without causing or suffering interferences. At the same time, the signal passes easily through walls, equipment and clothing [27-29] and more than one position can be tracked simultaneously. Moreover, RFID-UWB systems overcome limitations due to reflection, refraction, and diffraction phenomena, using pulses for the broadband transmission. The use of RFID-UWB offers other advantages, such as no line-of-sight requirements, high accuracy and resolution, lighter weight (the weight for each tag is less than 12 g) and the possibility to trace multiple resources at the same time, real-time and three-dimensionally. Furthermore, RFID-UWB sensors are cheaper, which make the RFID-UWB positioning system a cost-effective solution.

An RFID-UWB system comprises a computer and a hub (including a graphical interface), RFID-UWB sensors to record signals in real-time, RFID-UWB tags at low and high power and shielded CAT-5 cables. A set of sensors is positioned around the perimeter of the measured area. They receive pulses emitted by tags that include a set of data and are subsequently processed by the central hub.

The next section will describe in detail some experimental equipments developed by the authors based on the RFID-UWB system used in on-going research focused on real-time material flow traceability systems.

3. Experimental study

In this section, the experimental study about the traceability of material flows through IPS system based on RFID-UWB technology and its results are presented.

3.1. Components of the RFID-UWB system

The authors chose the RFID-UWB system, among IPS technologies since it is able to ensure the highest accuracy and precision in the measurements thanks to the combined use of AOA and TDOA techniques. The system comprises sensors, tags, and the software location platform, described below.

- *Sensors:* RFID-UWB sensors receive pulses from tags. Each sensor can determine the azimuth point and the arrival angulation thanks to the AOA technique. In this case, if only one sensor receives the signal, the system can determine the 2D location of the tag. Instead, if the signal is captured by more than one sensor, connected each other, it is also possible to find out the TDOA and obtain 3D location of tags. The configuration used reduces the infrastructure requisites, and consequently the costs, and guarantees high reliability and robustness of the system. The main characteristics of the sensors are:

  o *Reactivity in real-time*: each sensor maintains a constant frequency of 160 Hz, which means the tag can be seen every 6.25 ms by each sensor;

  o *Flexible installations*: this kind of infrastructure can be used for both small and large installations. Several sensors can be integrated in a unique system to monitor a big area and manage a large number of tags simultaneously;

  o *Synchronism:* in order to guarantee synchronism, the sensors are cabled with CAT-5 cables. A cell made up of several sensors is able to cover 10,000 m^2 of environment. In order to extend the covered area, the cells can be connected to each other;

  o *Bidirectional communication:* the sensors support bidirectional communication at 2.45 GHz. This allows the system to dynamically manage tags in an optimal way;

  o *Connectionsof sensors:* the sensors can be connected with standard Ethernet cables or through wireless adaptors, using pre-existing infrastructures like access point, switch Ethernet and CAT-5 wiring for communication between the sensors and the server;

○ *Ease of maintenance:* the sensors are managed in a remote way through TCP/IP protocols and standard Ethernet for communication and configuration.

Figure 4 shows the sensors used in the experimental application.

Figure 4. Sensors used in the experimental application (courtesy of Ubisense Group plc)

• *Tags:* these are small and robust devices worn by a person or attached to an object to be accurately located within an indoor environment. Tags transmit brief RFID-UWB pulses that are received by sensors and are used to determine their position. The use of RFID-UWB pulses ensures both high precision (approximately 15 cm) and great reliability in complex indoor environments, characterized by noises like reflection from walls or the presence of metallic objects in indoor environments. Each tag is made up of movement detectors for instantaneous activation, LEDs for identification and buttons for executing particular operations. The main characteristics of tags are:

○ *Precise localization:* the tag transmits RFID-UWB radio pulses, used by the localization system for defining the tag position within 15 cm. The precision of the system is also maintained in complex indoor environments thanks to RFID-UWB technology. In this way it is possible to obtain accurate information on 3D positions even when the tag is detected by only two sensors;

○ *Bidirectional communication:* tags use a dual-radio system in addition to the mono-directional RFID-UWB radio communication, used for the spatial detection. The capacity of bidirectional communication allows the system to dynamically manage the update rate of tags, control of LEDs and battery status;

○ *Flexible update rate:* the software platform allows the update rate of tags to be varied. If a tag moves quickly, it can have high upgrading for more precise localization; instead, if it moves slowly the update rate could be reduced in order to save the battery. When the tag is at rest, it is put into energy saving mode thanks to a built-in motion sensor that allows restart in case of movement;

○ *Interactivebuttons:* slim tags have two buttons (while compact tags have only one button) to allow context-sensitive inputs in systems requiring interactivity. The applications can use tag localization to work according to the events. The application can send feedback to the user through LEDs or acoustic signals;

- ○ *Resistant and suitable*: tags are resistant in critical industrial environments, since they can withstand dust and water. They can also be installed in mechanical and electronic instrumentation safely;

- ○ *Battery life*: the techniques of low-consumption and power management affect the duration of the battery. In a typical application, in which a tag is used to identify an operator every 3 sec, the battery has an average duration of four years.

Figure 5 shows an example of compact tags (on the left) and slim tags (on the right).

Figure 5. Compact tags (on the left) and slim tags (on the right), used in the experimental application (courtesy of Ubisense Group plc)

- *Software Location Platform* is used to control and calibrate the system, to manage the locations of data generated by tags and received by sensors and to analyse, communicate and inform users on the data system. The software platform is made up of the *Location Engine Calibration* and *Location Platform*.

- *Location Platform* is a software that collects and processes data from sensors and tags, viewable thanks to a graphical interface. In this way, it is possible to obtain 2D and 3D maps of the environment and detected assets. The collected data can be sent to other systems for further analysis and stored within the platform to act as a database.

- *Location EngineCalibration (LEC)* allows the sensors to be set, calibrated, and configured in cells using a graphical user interface. It is designed to allow the simple coordination of data from sensors and tags in order to be integrated in other applications. The Location Engine is the base component of the software platform since it allows the creation and loading of maps, single cell creation and setup of tags and sensors (deciding master and slave sensors), and the calibration of the system sensitivity (fixing the "noise threshold").

- The Location Engine supports several algorithms to determine tag position through sensor measurements. Each algorithm has a set of parameters that regulate tags behaviour. These parameters are called *filters* and can be applied to a single tag or a group of tags. The Location Engine presents one algorithm without a filter and another four filtered algorithms:

- *No filtering algorithm*: in this configuration, no filters are applied. This means that the position is evaluated only by measuring AOA and TDOA at a specific moment. In this way, any previous data is not processed and the path and speed of movements are not considered. Not using filters does not allow optimal measurements to be obtained.

- Filtered algorithms try to interpret tag movements to predict their positions during further measurements. Information coming from AOA and TDOA techniques is analysed and compared with the expected position that will be used in further measurement. The filter can eliminate measurements that can be deteriorated by reflections or disturbed by external noises. In order to do so, it is necessary to identify a movement pattern for the filter that defines the limitations to which the measured object has to be subjected. The higher the number of applied limitations, the better the robustness of the measurement. The filtered algorithms are presented below:

- *Information filter*: the tag can move along three directions but, if it is not seen for a period, the movement pattern assumes that it is continuing to move according to the last speed value and along the last detected direction. This algorithm is used for assets that move with predictable speed and without direction limitations;

- *Fixed height information filter*: the tag is free to move horizontally, but the vertical movements have to remain close to a predetermined threshold height. In this case, if contact with the tag is lost, it is assumed that it continues to move with equal speed along the horizontal direction, remaining close to the vertical predetermined height. Like the previous algorithm, the level of uncertainty of the location increases with the time. This algorithm is mainly used for vehicles moving at high speed and in two directions;

- *Static information filter*: the tag is free to move in three directions. If the tag is not detected, its position is identified with the last one and the level of uncertainty of localization increases with the time. This algorithm is used for assets that do not normally move or move in an unpredictable way, such as operators. The algorithm does not have any spatial limitations, allowing the detection of 3D movements (for example the movement of people climbing the stairs);

- *Static fixed height information filter:* the tag is free to move horizontally, but it is limited to the vertical direction. If the tag is not seen, it is assumed that its position is the last one detected and the height is close to the prefixed limit. This algorithm is used for targets that do not normally move or move in unpredictable way. Because of its vertical limitation, it is used for vehicles, tools, and people that move in two dimensions.

The parameters that can be regulated by the filtered algorithms are:

- *Handover stickiness*: indicates the tag's adherence to the cell in which it is located. It is measured indicating the maximum number of failed measurements of the tag's position before considering it out of the cell;

- *Handover minimum sensor count*: describes the minimum number of sensors belonging to the cell;

- *Low support reset count*: defines the time in which the tag can be seen with low support modality (the situation in which the measurements rejected by the filter are more than the valid ones) before the reset of the filter;

- *Min reset measurements*: indicates the minimum number of support measurements before the reset of the filter;

- *Tag power class*: the filter can validate the tag's position based on the level of signal power received by each sensor. The filter has to recognize the type of tag that sends the signal, so as to interpret the received power correctly. The value 0 disables the function, value 1 indicates a compact standard tag and value 2 indicates a tag with amplified signal power;

- *Static distance*: describes the minimum distance travelled by a tag compared with the last one;

- *Static alpha*: defines the fraction (0.0 – 0.1) of the current measurement used by the filter when the tag is considered stable. The tag position is computed as follows:

- (alpha * current position) + (1.0 – alpha) * (last position)

- If alpha is close to 0.0, the movement of the tag will be significantly damped;

- *Max position variance*: describes the maximum variation in estimating the position;

- *Max valid position variance*: identifies the maximum variance in estimating the position. This value has to be less than or equal to the "max position variance". The difference between these two parameters is that if the uncertainty is higher than the max position variance, the forecast of the next localization does not change; while if the variance is higher than the max valid position variance, the filter will continue to track the position, but this measurement is not considered valid.

No-static algorithms can regulate other parameters such as:

- *Max velocity*: identifies the maximum velocity at which the object can move;

- *Horizontal velocity standard deviation*: the filter operates with a model of movement in which the tag velocity is considered constant. This parameter indicates the rate of velocity increase in X and Y as the time varies;

- *Vertical velocity standard deviation*: similar to the horizontal velocity standard deviation, but for the vertical velocity;

- *Vertical position standard deviation* (only for the filter with prefixed height): although the height is fixed, this parameter allows the tag to be varied along the vertical movement. If the value of this parameter is 0, the tag will only be detected in two dimensions;

- *Tag height above cell floor* (only for filter with prefixed height): fixes the value of height Z where the tag should always be.

Static algorithms can also regulate another parameter:

- *Horizontal position standard deviation*: the filter operates according to the movement pattern in which the tag's position is considered constant. The uncertainty of the tag's position increases with the time although the tag's forecasting continues to be in the last position. This parameter identifies the increasing rate of standard deviation of position in X and Y as the time varies.

It is possible to underline the difference between static and dynamic filtering algorithms. In the case of dynamic filter, there are long straight lines that identify the moments in which the sensors lose track of the tag and find it again few moments later. Consequently, the measurement's accuracy is low, mainly in the computing of distances travelled, which may be compromised. In the case of static filter, the traced path is very close to the real one, without straight lines, since the tag is always under control. Figure 6 shows an example of tracking of the same path using a dynamic filtering algorithm (on the left) and a static filtering algorithm (on the right).

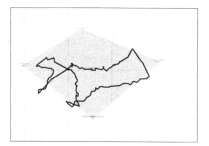 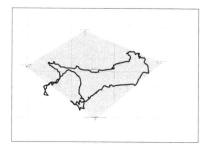

Figure 6. Path traced with dynamic filtering algorithm (on the left) and static filtering algorithm (on the right)

3.2. Installation and calibration of the system

The authors decided to install an IPS experimental system based on RFID-UWB technology in the Laboratory of Manufacturing System of Bologna University that, thanks to the presence of walls, machinery and metal objects, could be representative of a real industrial application.

Figure 7 shows the 2D map (on the left) and the 3D map (on the right) (obtained by LEC platform) of the laboratory, where the white squares indicate the position of the sensors. The optimal configuration needs sensors to be installed in the four corners of the building, but in actual fact, because of the presence of obstacles in the corners of the laboratory, the sensors are installed according to a rhombus distribution, able to guarantee total coverage of the area.

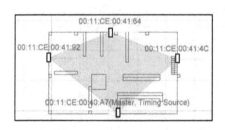 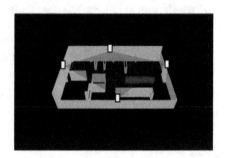

Figure 7. Map (on the left) and map (on the right) of the indoor environment considered in the application

The coordinates of sensors are presented in Table 1:

Sensors name	X [m]	Y [m]	Z [m]
00:11:CE:00:40:A7 (master)	15.618	-0.582	4.336
00:11:CE:00:41:4C (slave)	30.868	11.945	4.545
00:11:CE:00:41:64 (slave)	13.085	18.898	4.336
00:11:CE:00:41:92 (slave)	-0.308	11.039	4.651
STA (reference point)	15.409	10.833	2.100

Table 1. Coordinates of sensors

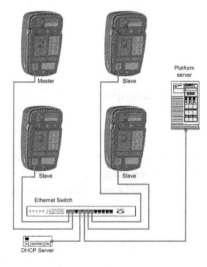

Figure 8. Connection of sensors with the system

The sensors have to be located as close as possible to the ceiling of the building to guarantee maximum coverage of the space and their angulation has to be directed towards the centre of the building. The sensors are grouped into rectangular cells, where they are connected to the switch POE that guarantees the power that is in turn linked with the PC (Figure 8). Each cell is characterized by a main sensor (master) that coordinates the activities of the other sensors (slave) and communicates with the tags. The master sensor has to be connected with the slaves by CAT-5 cables (Figure 9), in order to ensure the time synchronization. When the connection is made, the Location Engine Configurator is set to "Running" mode and the system is ready to work.

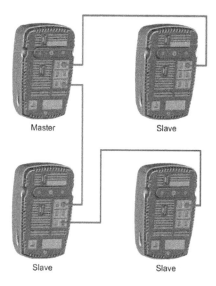

Master   Slave

Slave   Slave

Figure 9. Connection between master and slave sensors

The threshold level of the "background noise" has to be decided, so to allow the system to distinguish valid signals from environmental noises. In order to calibrate the sensors, the power level detected by them is measured, verifying that the "background noise" remains below the threshold level. After that, it is possible to calibrate the sensor orientation. The sensors are oriented to a known tag, taken as reference. Figure 10 shows the sensor calibration through AOA. The green lines connect each sensor to the detected position of the tag.

In order to activate the localization through TDOA, it is necessary to calibrate cables that synchronize all the slave sensors with the master. When the cable calibration is completed, blue strips are added to the green lines, one for each pair of sensors. In absence of obstacles, assets, and reflection phenomena, the blue lines are straight; in actual fact, they are curved lines, with increased bending as interferences and noises increase (Figure 11).

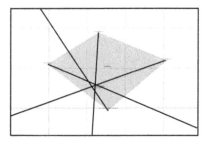

Figure 10. Calibration of sensors

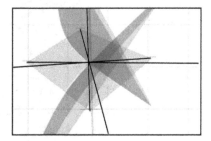

Figure 11. Calibration of cables

The system has to be connected with the layout of the environment to be monitored. A map of the laboratory has to be created, according to the external and internal walls, the columns and any other architecture present in the laboratory (Figure 12).

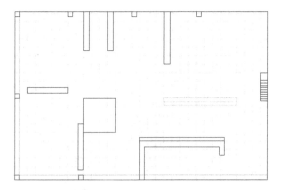

Figure 12. Map of the area controlled by the proposed system

When the map is loaded into the system, the coordinates of the sensors' position and some reference points within the area to be monitored have to be determined. A corner of the building is identified as the axis origin and is indicated by (0;0;0); the other corners will be identified with (X;0;0) and (0;Y;0) where X and Y are the length of the building sides. The level of floor is set as Z=0 so to use the 3D localization capacity of the system. In order to connect the position of the sensors with the laboratory's corner coordinates, the object localizations have to be calibrated, using known points as references. After that, the software will provide a 3D image of the area to be monitored.

In order to complete the calibration and verify the absence of errors, it is important to test the system, moving a tag within the area and ensuring that the sensors work correctly and that all necessary data is displayed.

3.3. Experimental evidence

The experimental research consists of several tests, static and dynamic.

The static tests consist of the identification of different points (to which tags are applied) within the area to be monitored. The sensors have to detect the coordinates of the tags to compare the estimated and detected coordinates of every point.

The dynamic tests consist of the application of a tag to an operator that goes around the monitored area. The operator follows prefixed paths, and the route and distance travelled by him are compared with the estimated values, measured in advance.

3.3.1. Static tests

In order to undertake the accuracy and precision of the proposed RFID-UWB system, the first test is the measurement of known point coordinates through a laser. 16 points within the monitored area, chosen according to the characteristics of visibility, proximity to metal objects and position, are identified.

The static tests are performed according to the variation of some tag parameters, such as:

- Filter used (*No-filter, Information Filter, Fixed Height Information Filter, Static Information Filter, Static Fixed Height Information Filter*);

- Update each four time slot;

- Frequency: 37 Hz;

- All tests are performed by putting the asset on a support 0.5 m high, except point 13, which is placed 2 m high.

Figure 13 shows the considered 16 points represented in the map of the laboratory.

For each point, four tests are performed in order to understand the average error between the estimated and detected coordinates.

Test 1

- Filter: *Static Fixed Height Information Filter;*
- Update each four time slot;
- Frequency: 37 Hz.

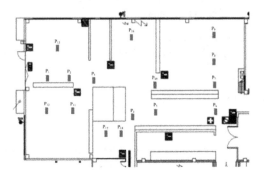

Figure 13. Reference points for static tests

Table 2 presents the estimated and detected coordinates of the 16 points, specifying the error between them.

Point	X [m]	Y [m]	X detected [m]	Y detected [m]	Error [m]
1	5.826	11.113	6.1007	11.0112	0.2930
2	7.976	10.991	7.6250	10.7717	0.4138
3	11.389	11.207	11.5024	11.0014	0.2347
4	15.693	7.138	15.9791	7.2085	0.2946
5	18.627	7.138	18.6374	7.1973	0.0602
6	26.39	7.138	26.1077	7.4651	0.4320
7	26.39	11.204	26.5720	10.8224	0.4227
8	26.39	14.028	26.2713	13.4417	0.5980
9	26.39	16.6	29.1303	18.6123	3.3998
10	19.347	11.207	19.5315	10.5318	0.6999
11	8.17	7.113	7.8618	7.2394	0.3330
12	3.415	7.05	3.7401	8.0534	1.0548
13	3.399	15.739	3.3871	14.3100	1.4289
14	16.397	2.748	14.9795	3.2646	1.5085
15	11.367	2.748	11.7532	4.3628	1.6602
16	15.637	15.187	15.8656	14.9317	0.3426

Table 2. Analysis of static *Test 1*

The average error of *Test 1* is 0.8236 m.

The points situated in the best positions (excluding points 9, 12, 13, 14 and 15) are located with high accuracy and present an average error of 37 cm. The worst result of point 9 is due to the presence of numerous obstacles around the considered area that make the tag visible to only one sensor. The same causes also influence the detection of points 12, 13, 14 and 15, although with lower impact.

Test 2

• Filter: any filter applied;

• Update each four time slot;

• Frequency: 37 Hz.

Table 3 presents the estimated and detected coordinates of the 16 points, specifying the error between them.

Point	X [m]	Y [m]	X detected [m]	Y detected [m]	Error [m]
1	5.826	11.113	5.8597	10.9974	0.1203
2	7.976	10.991	7.9914	11.1263	0.1362
3	11.389	11.207	11.4886	11.1252	0.1289
4	15.693	7.138	15.8387	7.2311	0.1729
5	18.627	7.138	18.4915	7.0736	0.1499
6	26.39	7.138	21.3617	3.6867	6.0987
7	26.39	11.204	28.4824	11.4226	2.1038
8	26.39	14.028	25.6281	13.6126	0.8676
9	26.39	16.6	28.6270	9.2454	7.6872
10	19.347	11.207	19.4776	10.3303	0.8863
11	8.17	7.113	7.7177	7.6014	0.6656
12	3.415	7.05	6.5425	5.6330	3.4335
13	3.399	15.739	3.5185	14.9626	0.7855
14	16.397	2.748	14.7514	2.9406	1.6567
15	11.367	2.748	10.9230	4.2645	1.5800
16	15.637	15.187	15.7053	15.1101	0.1028

Table 3. Analysis of static *Test 2*

The average error of *Test 2* is 1.661 m.

The absence of filters means that the oscillations of the tag positions are not damped. This leads to the worst result of all the tests. It is possible to note that the easily reachable and visible points present low error values, while for the most critical points the system performance is worse, even reaching high error values (in the order of metres).

Test 3

- Filter: *Static Information Filter;*
- Update each four time slot;
- Frequency: 37 Hz.

Table 4 presents the estimated and detected coordinates of the 16 points, specifying the error between them.

Point	X [m]	Y [m]	X detected [m]	Y detected [m]	Error [m]
1	5.826	11.113	6.0484	10.8824	0.3203
2	7.976	10.991	7.4876	11.0006	0.4884
3	11.389	11.207	11.3885	10.9646	0.2423
4	15.693	7.138	15.9165	6.9975	0.2640
5	18.627	7.138	18.6268	7.0524	0.0855
6	26.39	7.138	21.5961	4.4598	5.4912
7	26.39	11.204	26.5539	10.9868	0.2720
8	26.39	14.028	25.7937	13.2303	0.9959
9	26.39	16.6	22.0670	9.2817	8.4996
10	19.347	11.207	20.3456	10.0280	1.5450
11	8.17	7.113	7.1760	7.0652	0.9950
12	3.415	7.05	3.4986	7.7185	0.6737
13	3.399	15.739	3.1959	14.0444	1.7067
14	16.397	2.7481	14.9009	2.8403	1.4989
15	11.367	2.7481	12.5534	3.8802	1.6399
16	15.637	15.187	15.8690	14.8410	0.4165

Table 4. Analysis of static *Test 3*

The average error of *Test 3* is 1.5709 m.

Like the other two tests, points 6 and 9 present largely incorrect values, because of the condition of the area in which they are located. The other values are in line with the estimated measurements.

Test 4

- Filter: *Information Filter;*
- Update each four time slot;
- Frequency: 37 Hz.

Table 5 presents the estimated and detected coordinates of the 16 points, specifying the error between them.

Point	X [m]	Y [m]	X detected [m]	Y detected [m]	Error [m]
1	5.826	11.113	6.1134	10.8474	0.3913
2	7.976	10.991	7.4098	11.1165	0.5798
3	11.389	11.207	11.4637	11.0003	0.2197
4	15.693	7.138	15.7641	7.0616	0.1043
5	18.627	7.138	18.7090	7.1121	0.0860
6	26.39	7.138	20.3066	4.3934	6.6738
7	26.39	11.204	26.5290	10.9065	0.3283
8	26.39	14.028	22.7547	9.2591	5.9963
9	26.39	16.6	27.3837	17.1635	1.1424
10	19.347	11.207	19.5325	10.7986	0.4484
11	8.17	7.113	5.5139	7.2697	2.6607
12	3.415	7.05	4.0360	7.3234	0.6786
13	3.399	15.739	3.4861	14.5849	1.1573
14	16.397	2.7481	14.9694	3.0857	1.4668
15	11.367	2.7481	11.8527	3.4522	0.8554
16	15.637	15.187	15.7347	15.1152	0.1212

Table 5. Analysis of static *Test 4*

The average error of *Test 4* is 1.4319 m.

In this case, the results are better than *Test 2* and *Test 3*, but the problems regarding the presence of obstacles in the area to be monitored, noted during the other tests, remain.

From a comparison between the four static tests (Table 6), it is possible to note that the best algorithm in terms of the lowest average error between estimated and detected tag position is *Test 1* that uses a *Static Fixed Height Information Filter*.

Filter used	Average error [m]
Static Fixed Height Information Filter	0.8236
Any filter applied	1.661
Static Information Filter	1.5709
Information Filter	1.4319

Table 6. Comparison between the average errors of static tests

3.3.2. Dynamic tests

Dynamic tests are performed by applying a tag to an operator that goes around the laboratory following prefixed paths. The length of these paths, measured in advance, is compared with the real distance travelled by the operator. In this way, it is possible to see the precision of each known point and test the capacity of the system to reconstruct the trajectory.

The first part of the paragraph presents the results obtained by dynamic tests, using a static filter (*Static Information Filter*), while the second part shows the same results using a dynamic filter (*Information Filter*), underlining the differences between them.

3.3.2.1. Dynamic tests using Static Information Filter

Four tests are performed, according to the following parameters:

- Filter: *Static Information Filter*;
- Update each four time slot;
- Frequency: 37 Hz;
- Threshold speed: 2 m/sec;
- Velocity of tag: 2 m/sec at a constant height of 1.5 m.

In order to cover the whole interested area, several proof paths are decided and measured in advance.

Test 1

The path is 28.8 m long: the first part is made up of an area with good coverage by sensors without obstacles, while in the second part the operator has to cross an area with numerous obstacles and metallic materials. Figure 14 shows the estimated path (Figure 14a) and the detected path travelled by the operator, obtained using LEC software (Figure 14b).

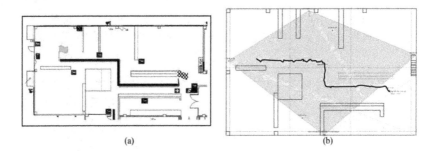

(a)   (b)

Figure 14. a. Estimated path of dynamic *Test 1* b. Detected path of dynamic *Test 1*

Table 7 shows the detected and measured distances and the difference between them.

Distance estimated [m]	Distance travelled [m]	Error [m]	Error [%]
28.8	31.22	2.421	8.408

Table 7. Synthesis of dynamic *Test 1*

Test 2

The path is 30 m long and it travels around a metallic shelf in the centre of the laboratory. Figure 15 shows the estimated path (Figure 15a) and the detected path travelled by the operator, obtained using LEC software (Figure 15b). As can be seen from Figure 15b, the blue-sky line representing the path, presents some noises, due to the momentary loss of the signal.

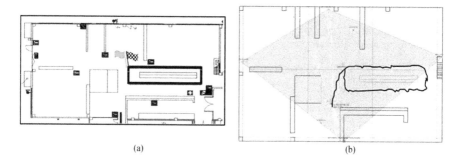

(a)   (b)

Figure 15. a. Estimated path of dynamic *Test 2* b. Detected path of dynamic *Test 2*

Table 8 shows the detected and measured distances and the difference between them.

Distance estimated [m]	Distance travelled [m]	Error [m]	Error [%]
30	30.45	0.4594	1.5315

Table 8. Synthesis of dynamic *Test 2*

Test 3

The path is 23.5 m long: the first part is made up of an area with low coverage, because of the presence of walls, shelves and several metallic machines and objects. In the final part, the path is made up of an area surrounded by machineries and this makes the correct localization of the tag difficult. Figure 16 shows the estimated path (Figure 16a) and the detected path travelled by the operator, obtained using LEC software (Figure 16b).

Table 9 shows the detected and measured distances and the difference between them.

Distance estimated [m]	Distance travelled [m]	Error [m]	Error [%]
23.5	24.02	0.5199	2.2125

Table 9. Synthesis of dynamic *Test 3*

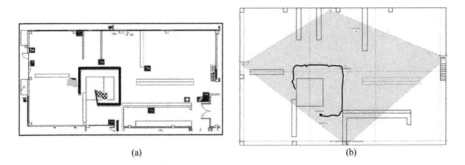

(a)   (b)

Figure 16. a. Estimated path of dynamic *Test 3* b. Detected path of dynamic *Test 3*

Test 4

The path is 11.5 m long. It is situated in a complex environment, characterized by the presence of walls and several machines that strongly hinder correct signal reception by the sensors. Indeed, it is possible to observe the irregular trend that causes problems in the correct evaluation of the distance travelled. Figure 17 shows the estimated path (Figure 17a) and the detected path travelled by the operator, obtained using LEC software (Figure 17b).

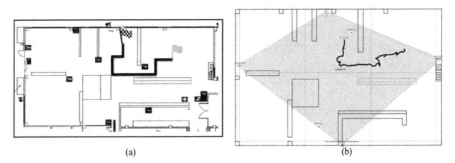

(a)   (b)

Figure 17. a. Estimated path of dynamic *Test 4* b. Detected path of dynamic *Test 4*

Table 10 shows the detected and measured distances and the difference between them.

Distance estimated [m]	Distance travelled [m]	Error [m]	Error [%]
11.5	14.42	2.9285	25.465

Table 10. Synthesis of dynamic *Test 4*

3.3.2.2. Dynamic tests with Information Filter

The authors decide to re-apply the same tests applying a dynamic filter, called *Information Filter*, to the algorithm, in order to compare the results with those obtained by using a static filter. If the sensors lose the signal, the static filter maintains the last detected position and updates it when a valid signal arrives. The dynamic filter, instead, stores the velocity and the direction of the tag moment all times and, in case of absence of valid signals, it assumes that the target continues to move in the same direction and at the same velocity as the last measurement. The use of a dynamic filter results in lower performance of operations for the reconstruction of trajectories, since the paths do not reflect the real tag movements.

The tests are performed according to the same parameters as the dynamic tests with a static filter:

- Filter: *Information Filter*;
- Update each four time slot;
- Frequency: 37 Hz;
- Threshold speed: 2 m/sec;
- Velocity of tag: 2 m/sec at a constant height of 1.5m.

The paths are the same as the dynamic tests with static filter.

Test 1

The application of a dynamic filter does not heavily modify the results, except for the central stretch and the last part of the path, since it is made up of metallic materials. Figure 18 shows the comparison between the maps obtained using LEC software, in the case of static (Figure 18a) and dynamic filter (Figure 18b). The arrows show the main differences between the paths travelled using a static and a dynamic filter.

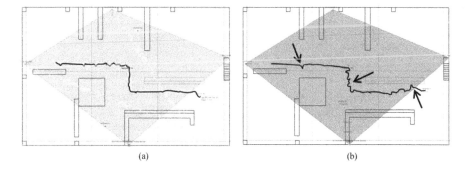

(a)   (b)

Figure 18. a. Path travelled using static filter b. Path travelled using dynamic filter

The path travelled by an operator in the laboratory using a dynamic filter, presents more noise than that travelled using a static filter. It is possible to note some peaks along the path, due to loss of signal. Indeed the *Information Filter* allows the target to move along the three dimensions, but, if it is not seen for a period, the system assumes that it is moving along the same direction and at the same velocity.

Test 2

In this case, the path is strongly modified at the point where the signal is lost. In particular, it is possible to observe the formation of straight lines that indicate that sensors were not able to detect the tag presence for some seconds. In this way, the last trajectory is maintained, but it does not reflect the real path travelled by the target. Figure 19 shows the comparison between the maps obtained using LEC software, in the case of static (Figure 19a) and dynamic filter (Figure 19b). The arrows show straight lines formed because of the loss of signal by the sensors, unlike the case of a static filter.

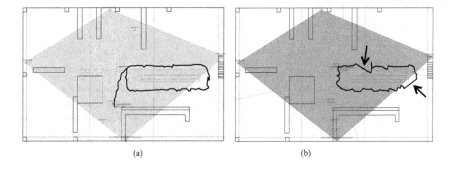

(a) (b)

Figure 19. a. Path travelled using static filter b. Path travelled using dynamic filter

Test 3

In this case, the errors in the traceability of the path are less evident than in the last case, but it is possible to note that the line appears more indented. This is an indication of more noises during localization. Moreover, in the final part, the trace overlaps with a wall, underlining the limits of the localization with the dynamic filter. Figure 20 shows the comparison between the maps obtained using LEC software, in the case of static (Figure 20a) and dynamic filter (Figure 20b). The arrows show the main differences between the paths travelled using a static and a dynamic filter.

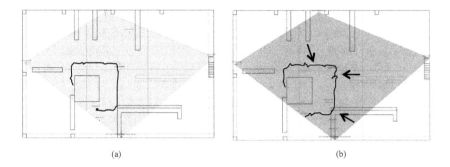

(a)   (b)

Figure 20. a. Path travelled using static filter b. Path travelled using dynamic filter

Test 4

In this case, the errors in the traceability of the path are evident, because of the critical environment in which the path is travelled. In the middle of the path the signal is lost and found again only in the proximity of the final part of the path. This leads to the creation of a straight line that does not reflect the real movement of the tag. Figure 21 shows the comparison between the maps obtained using LEC software, in the case of static (Figure 21a) and dynamic filter (Figure 21b). The arrows show the main differences between the paths travelled using a static and a dynamic filter.

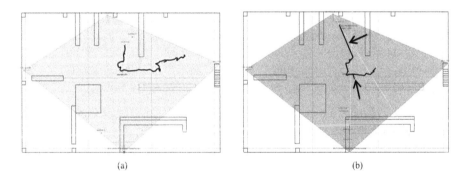

(a)   (b)

Figure 21. a. Path travelled using static filter b. Path travelled using dynamic filter

The algorithm using a static filter provides better results than that using the dynamic filter. A comparisons between the two algorithms show that if the sensors lose the tag signals for a period, the system assumes that the tags continue to move according to the last velocity value and along the last direction of movement. The greater the moment of no-detection of tag's position, the higher the inaccuracy of the system, causing a distortion of the path.

3.4. RFID technology applied to packaging system

RFID technology is introduced in the packaging sector due to the logistics advantages regarding the utilization of automatic identification systems. This introduction mainly focuses on the secondary and tertiary packaging levels because the utilization in the item level (product identification) has been difficult to justify in economic terms [30]. Specifically, 250-300 millions of tags were used in 2006 in the tertiary level [31]. Furthermore, Thoroe et al. [32] have predicted that in 2016 there will be 450 times more RFID tags in use than today. Therefore, a rapid increase in RFID tag consumption is expected in the packaging sector.

Technological developments in recent years, along with a reduction in tag price and emerging standards have facilitated trials and rollouts of RFID technology in packaging. A study conducted by IDTechEx Limited [33] stated that the main benefits of RFID technology in packaging are better service and lower costs.

Packaging incorporating RFID technology is usually referred to as *smart packaging* (called also *active* or *intelligent packaging*) and it is commonly used to describe packaging with different types of value-adding technologies, for example placing in the package a smart label or tag. The term smart packaging was used by Yam [34] in 1999 to emphasize the role of packaging as an intelligent messenger or an information link. According to the Smart Packaging Journal [35], smart packaging is described as *packaging that employs features of high added value that enhances the functionality of the product* and its core is responsive features. These high-value features have a variety of characteristics, but are mainly made up of mechanical or electronic technology features such as mechanical medicines, dispenser of packaging tagged with electronic devices like RFID technology. Smart packaging is often used to refer to electronic responsive features where data is electronically sensed on the package from a distance, using an automatic identification system as the RFID technology. Schilthuizen [36] pointed out that identification and sensor technology enable intelligent functions in packaging.

Usually packages – and the products contained within them – are traced with systems obtained through asynchronous fulfilment of doorways by materials. In such cases, the tracking is totally manual, executed by operators. These manual activities could be eliminated or replaced by an automated identification activity, using an RFID system. The application of RFID to packaging allows more frequent and automated identification of packages (e.g. pallets, cases, and items) increasing the accuracy of the system, reducing the labour and time needed to perform the identification of packages and enabling near real-time visibility, which in turn facilities the coordination of activities within and between processes. The costs of RFID technology in packaging and potential benefits will vary, according to the packaging level that is tagged. Figure 22 (modified version of [25]) illustrates the influence that tagging different packaging levels has on the retail supply chain processes.

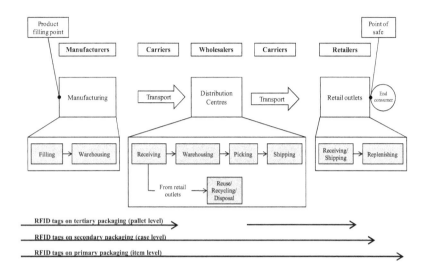

Figure 22. The influence of tagging different packaging levels along the supply chain (modified version of [25])

As can be noted in Figure 22, RFID tags on tertiary packaging may be used from the filling to the storing process. Furthermore, the tags on tertiary packaging may be used from the shipping process of the distribution centre to the receiving and shipping process of the retail outlet. RFID tags on secondary packaging could be used further downstream in the supply chain than tagged tertiary packaging, i.e. from the filling process and all the way to the replenishing process. Irrespective of the activities within the replenishment processes, tagging of primary packaging may be used in the whole supply chain, from the point of filling by the manufacturer to the point of sale in the retail outlet. Tagging of primary and secondary packaging could also provide opportunities beyond the point of sale in retail outlets e.g. recycling, reusing, and post-sales service and support. Although tagging on the primary packaging level will bring about the greatest level of benefits for the retail supply chain, tagging on secondary and tertiary packaging levels could provide valuable benefits for the supply chain. The model presented in Figure 22 indicates that a manufacturer who applies the tags to packaging can gain direct benefits from primary and secondary packaging tagging. According to [25], the average time to pick an order decreases by roughly 25% when RFID technology is used in secondary packaging. This means that the workforce conducting the picking activity, which is the core and the most labour-intensive activity in distribution centres could be reduced by approximately 25%. Hellström [25] also stated that the ability to automatically generate orders by capturing the inventory levels through tagging of primary packaging could reduce out-of-stock situations by approximately 50%.

Figure 23 shows the traceability of a primary package patterns within a manufacturing company (in particular in an assembly station) using the RFID-UWB system.

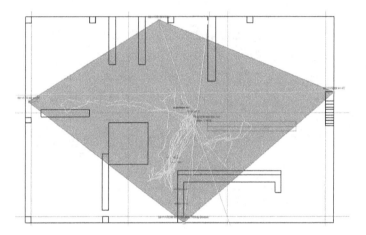

Figure 23. Traceability of a package with RFID-UWB system (Spaghetti Chart)

The traditional approach provides the well known *Spaghetti Chart* (manually realized). In addition, the data is approximate and does not provide precise values. In order to overcome the difficulty in analysing data from a traditional tracking method, Real Time Location System is perfect for the traceability of goods.

The framework on RFID and packaging shows the importance of tracing packages since several benefits can be achieved (e.g. reduction of costs and time, increase of efficiency and effectiveness, accuracy of the activities along the supply chain, security of the products, etc.). The RFID-UWB system presented in the chapter is perfectly aligned with the problem of package traceability.

3.5. Results and discussion

The results obtained during the static tests show that the average error between the estimated and the detected measurements is approximately 1 m. However, it is important to consider the non-optimal installation of the sensors. In fact, the most suitable arrangement to obtain the maximum coverage of the area is obtained by placing the sensors at the corners of the area to be monitored. Because of the presence of obstacles and metallic objects, the authors have had to opt for an alternative solution, placing each sensor in the middle of each wall. This causes the incomplete coverage of the monitored area.

Despite this limitation, the authors have chosen to include in the tests some points located outside the optimal coverage area. In these cases, the obtained accuracy is much lower than that obtained by points located where the coverage is maximum. For example, in some cases, there are errors of several meters, not compatible with the project needs and that affect the estimation of the average error that increases greatly. For this reason, these points are eliminated in the computing of the average error estimation, obtaining a considerable improvement in the accuracy, reaching an average error of 40 cm.

Regarding dynamic tests, the problems connected with the layout of the area are the same as the static ones. The authors have set the tests to simulate paths all around the area. Several critical points cannot be seen by the sensors. In particular, in some areas, the tag is seen only by one or two sensors, which results in inaccuracies in the traceability of the path travelled by the target.

Unlike the static tests, during the dynamic tests, it is necessary to control the typology of filters used. The results show that the best performance is obtained using a static (*Static Information Filter*), rather than a dynamic (*Information Filter*) filter, with an error of 5% between the estimated and the real distance travelled. If the sensors lose the signal and the filter is dynamic, the system continues to see the tag moving along the same direction and at the same velocity as the previous measurement. On the contrary, if the sensors do not see the signal and the filter is static, the system assumes that the position of tag is the same as the last measured. In conclusion, systems using dynamic filters provide less accurate results than systems using static filters.

In order to improve the performance of the system, several changes could be made:

- Install sensors according to the optimal layout, locating them in the corners of the monitored area, so to obtain greater coverage and eliminate points in which the intensity of the signal is low;

- Locate sensors as high as possible so that each point is in the line-of-sight of at least two sensors;

- Customize the filter configuration, finding a combination of parameters, better suited to the characteristics of the monitored area, and type of application (velocity of movement, static or dynamic detection, etc.) to be achieved.

4. Conclusion

In recent years, more and more companies are recognizing the importance of tracing logistics flows in indoor environments (e.g. factories, warehouses, production plants, etc.). One of the best ways to analyse internal flows of materials is the Real Time Location System (RTLS) and in particular the Indoor Positioning System (IPS). IPS is a process that continuously determines in real-time the position of something or someone in a physical space [9]. RFID-UWB (Radio Frequency IDentification-Ultra Wide Band) technology is the best method to use for tracing targets within a company, among others. The main advantages of RFID-UWB technology are that it requires a very low level of power and can be used in close proximity to other RF signals without causing or suffering interferences. At the same time, the signal passes easily through walls, equipment, and clothing [27-29] and more than one position can be tracked simultaneously. The use of RFID-UWB offers other advantages, such as no line-of-sight requirements, high accuracy and resolution and the possibility to trace multiple resources in real-time. Furthermore, RFID-UWB sensors are cheaper, and this makes the RFID-UWB positioning system a cost-effective solution.

In order to trace the position and to map the movements of targets (e.g. people, materials, products, vehicles, information), the authors have developed an experimental IPS system based on RFID-UWB technology in the Laboratory of Manufacturing System of Bologna University which, thanks to the presence of walls, machineries and metal objects, can represent a real industrial application. The system is made up of active tags – positioned on fork-lifts, packages, or operators –, sensors that receive the signal from tags, and a software platform that collects data in order to present, analyse and communicate information to the final customer. The tags, which must be positioned around the tested areas, transmit short pulses to the sensors, organized in rectangular cells. Each cell is characterized by a main sensor (*master*) that coordinates the activities of the other sensors (*slave*) and communicates to the tags the detected position within the cell. The software platform carries out the positioning calculations based on information by the sensors and then analyses the results.

The experimental research consists of several tests, static and dynamic. The results present useful conclusions in terms of system performance, accuracy, and measurement precision.

The static tests give good results in terms of average error (approximately 40 cm) between the estimated and detected position of all considered points. The dynamic tests are performed using filters that regulate the behaviour of tags. The filters can be static or dynamic. The tests performed by applying a static filter produce better results compared with dynamic filter. If the sensors lose the signal and the filter is dynamic, the system continues to see the tag moving along the same direction and at the same velocity as the last measurement. On the other hand, if the filter is static, the system assumes that the position of tag is the same as the last measured. In conclusion, systems using static filter provide more accurate results (with an average error between the estimated and detected real distance travelled of 5%) than systems using the dynamic filter.

RFID technology can be also applied to packaging. Although the use of RFID technology in packaging is still limited, more and more companies are recognizing the importance of tracing packaging products moving within indoor environments. During recent decades, the importance of packaging and its functions is been increasing. Packaging is considered an integral element of logistics systems and its main function is to protect and preserve products. More often companies have to transport and distribute particular goods (e.g. dangerous or explosive products) or expensive products, such as some kinds of medicines. Since companies need to reduce thefts, increase security, and reduce costs and time spent on the traceability of products, they are starting to use RFID in packaging.

Rapid advances in factory automation in general and packaging operations in particular have posed a challenge for engineering and technology programs for educating a qualified workforce to design, operate and maintain cutting edge techniques such as RFID systems [37]. The system proposed by the authors tries to play this challenge.

Author details

Alberto Regattieri and Giulia Santarelli

DIN – Department of Industrial Engineering, University of Bologna, Bologna, Italy

References

[1] Choi T.Y., Krause D.R. The supply base and its complexity: implications for transactions costs, risks, responsiveness and innovation. Journal of Operations Management 2006; 24 637-652.

[2] Myerson J.M. RFID in the supply chain: a guide to selection and implementation. Auerbach Publications; 2007.

[3] European Parliament. Regulation (EC) No. 178/2002 of the European Parliament and of the Council of 28 January 2002 laying down the general principles and requirements of food law, establishing the European Food Safety Authority and laying down procedures in matters of food safety. Official Journal of the European Communities 2002; L 13, 1.2.2002, pp 1-24.

[4] Regattieri A., Gamberi M., Manzini R. Traceability of food products: general framework and experimental evidence. Journal of Food Engineering 2007; 81 347-356.

[5] Martínez-Sala A.S., Egea-López E., García-Sánchez F., García-Haro J. Tracking of returnable packaging and transport units with active RFID in the grocery supply chain. Computers in Industry 2009; 60(3) 161-171.

[6] Finkenzeller K. RFID-Handbook: Fundamentals and applications in contactless smart cards and identification. John Wiley & Sons, Ltd; 2003.

[7] Mahalik N.P. Processing and packaging automation systems: a review. Sensory and Instrumentation for Food Quality 2009; 3 12-25.

[8] Liu, H., Banerjee, P. and Liu, J. Survey of wireless indoor positioning techniques and systems. IEEE Transactions on Systems, Manufacturing and Cybernetics 2007; 37(6) 1067-1081.

[9] Hightower J. and Borriello G. Location systems for ubiquitous computing. Computer 2001; 24(8).

[10] Gu Y., Lo A., Niemegeers I. A survey of indoor positioning systems for wireless personal networks. IEEE Communications Systems & Tutorials 2009; 11(1) First quarter.

[11] Curran K., Furey E., Lunney T., Santos J. and Woods D. An evaluation of indoor location determination technologies. Intelligent Systems Research Centre, Faculty of Computing and Engineering, University of Ulster 2011.

[12] Vossiek M, Wiebking M., Gulden L., Weighardt P., Hoffman J. Wireless local positioning – Concepts, solutions, applications. Proceedings of Wireless Communications & Networking Conference, Boston, Massachusetts, August 2003.

[13] Casas R., Cuartielles D., Marco A., Gracia H.J. and Falc J.L. Hidden issues in deploying in indoor location system. IEEE Pervasive Computing 2007; 6(2) 62-69.

[14] Lee C., Chang Y., Park G., Ryu J., Jeong S. and Park S. Indoor positioning system based on incident angles of infrared emitters. Proceedings of 30th Annual Conference of the IEEE Industrial Electronics Society, Busan, Korea, November 2004.

[15] Zhang D., Xia F., Yang Z., Yao L., Zhao W. Localization technologies for indoor human tracking. IEEE Communications Surveys & Tutorials 2009; 11(1) First quarter.

[16] Lin T and Lin P. Performance comparison of indoor positioning techniques based on location fingerprinting in wireless networks. Proceedings of International Conference Wireless Network, Communications and Mobile Computing, Maui, Hawaii, June 2005.

[17] Ni L.M. and Liu Y. LANDMARC: Indoor location sensing using active RFID. Proceedings IEEE International Conference on Pervasive Computing and Communications, Forth Worth, Texas, 2003, pp.407-416.

[18] Wang Y, Jia X., Lee H.K. An indoor wireless positioning system based on wireless local area network infrastructure. Proceedings of the 6th International Symposium on Satellite Navigation Technology Including Mobile Positioning and Location Services, Melbourne, Australia, July 2003.

[19] Bruno R, Delmastro F. Design and analysis of a bluetooth-based indoor localization system. Proceedings of Personal Wireless Communication, Venezia, ITALY, November 2003.

[20] Niculescu D. University R. Positioning in ad hoc sensor networks. IEEE Network Magazine 2004; 18(4).

[21] Pruitt S. Wal-Mart begins RFID trial in Texas. Computerworld 2004.

[22] Collins J. Tesco begins RFID rollout. RFID Journal 2004.

[23] Collins J. Marks & Spencer to extend trial to 53 stores. RFID Journal 2005.

[24] RFID Radio. RFID's reduction of out-of-stock study at Wal-Mart. RFID Radio 2007.

[25] Hellström D. (2004). Exploring the potential of using radio frequency identification technology in retail supply chain – A packaging logistics perspective. PhD Thesis. Lund University, Sweden; 2004.

[26] Federal Communications Commission, FCC. Revision of part 15 of the commission's rules regarding ultra-wideband transmission systems. Report and order, FCC 02 48, April 2002.

[27] Gezici S., Tian Z., Giannakis G.V., Kobaysahi H., Molish A.F., Poor H.V. and Sahino-glu Z. Localization via ultra-wideband radios: a look at positioning aspects for future sensor networks. IEEE Signal Processing Magazine 2005; 22(4) 70-84.

[28] Fontana R.J. Recent system applications of short-pulse ultra-wideband (UWB) technology. IEEE Transactions on Microwave Theory and Techniques 2004; 52(9) 2087-2104.

[29] Molish A.F. Ultrawideband propagation channels, theory, measurement, and modelling. IEEE Transactions on Vehicular Technology 2005; 54(5) 1528-1545.

[30] Aliaga C., Ferreira B., Hortal M., Pancorbo M.A., López J.M., Navas F.J. Influence of RFID tags on recyclability of plastic packaging. Waste Management 2011; 31(6) 1133-1138.

[31] IDTechEx, 2006. Has the pallet/case market for RFID tags and other hardware become the nearest thing to a black hole in the RFID universe in 2006, thanks to reluctant mandated customers, technical problems and pricing for volumes that never came? www.Idteches.com.

[32] Thoroe L., Melski A., Schumann M. Item-level RFID: curse or blessing for recycling and waste management? Proceedings of the 42nd Hawaii International Conference on System Sciences, Hawaii, January 2009.

[33] IDTechEx Limited. Why the use of RFID in smart packaging? Smart Packaging Journal 2002; 1(2) 9-10.

[34] Yam K.L. Intelligent packaging for the future smart kitchen. Packaging Technology and Science 2000; 13 83-85.

[35] IDTechEx Limited. A day with smart packaging. Smart Packaging Journal 2002; 1(3) 1-19.

[36] Schilthuizen S.F. Communications with your packaging: possibilities for intelligent functions and identification methods in packaging. Packaging Technology and Science 1999; 12 225-228.

[37] Djassemi M., Singh J. The use of RFID in manufacturing and packaging technology laboratories. Proceedings of 3rd International Conference on Manufacturing Education. San Luis Obispo, California, June 2005.

Chapter 7

Possibility of RFID in Conditions of Postal Operators

Juraj Vaculík, Peter Kolarovszki and Jiří Tengler

Additional information is available at the end of the chapter

1. Introduction

Radio frequency identification is becoming a modern trend in many sectors. It provides a contactless identification, tracking and tracing of goods, property and people in real time. Increase efficiency, performance and competitiveness. One area of application of RFID technology is also postal processes. In this context there are several question of feasibility of the use of identification of letters. In addition to the costs associated with the introduction of technology is necessary to examine the feasibility of using RFID technology in the field of postal processes.

Today, postal operations have implemented RFID in various closed-loop systems to measure, monitor, and improve operations. For example, RFID is being used to monitor international mail service between major hubs. By randomly "seeding" tagged letters into trays, elapsed delivery time can be measured. This allows service issues to be identified and addressed in a reliable and cost-effective manner.

Other postal operations have piloted tracking mail containers to measure trailer utilization and to track container locations. Manual container tracking systems tend to break down when volumes are high and there's a deadline to meet departure times. By allowing information to be captured automatically, RFID makes sure it is done, even under stressful conditions. Postal managers can rely on the information to make decisions that improve transportation costs and to relocate containers when needed. RFID-tracked mailbags, which provide delivery status, have already been created for priority mail services. Tagged mailbags are automatically read at specific points in the network to provide this automated track-and-trace capability. Four additional areas can benefit from the cheap, accurate, and pervasive information obtained using RFID. Each of them has the prospect for returning substantial monetary benefits, as well as having the potential to significantly upgrade postal service capabilities, an ever more important consideration in the competitive delivery market.

Chapter is divided on several parts. We will be talk about basic of RFID, possibility of technology in postal and logistics processes, other mobile technology in processes, security of technology with contents to postal services, impact of operational characteristic on the readability and finally results of testing RFID technology in our laboratory of Automated identification and data capture (AIDC Lab) of University of Žilina.

2. Basic of RFID technology architecture

The RFID system architecture consists of a reader and a tag (also known as a label or chip). The reader queries the tag, obtains information, and then takes action based on that information. That action may display a number on a hand held device, or it may pass information on to a POS system, an inventory database or relay it to a backend payment system thousands of miles away. Let's looks at some of the basic components of a typical RFID system.

2.1. RFID tag/label

RFID units are in a class of radio devices known as transponders. A transponder is a combination transmitter and receiver, which is designed to receive a specific radio signal and automatically transmit a reply. Transponders used in RFID are commonly called tags, chips, or labels, which are fairly interchangeable, although "chip" implies a smaller unit, and "tag" is used for larger devices. The designator label is mainly used for the labels that contain an RFID device. Tags are categorized into four types based on the power source for communication and other functionality (Figure 1):

- A passive tag uses the electromagnetic energy it receives from an interrogator's transmission to reply to the interrogator. The reply signal from a passive tag, which is also known as the backscattered signal, has only a fraction of the power of the interrogator's signal. This limited power significantly restricts the operating range of the tag. Since passive tags are low power devices, they can only support data processing of limited complexity. On the other hand, passive tags typically are cheaper, smaller, and lighter than other types of tags, which are compelling advantages for many RFID applications. [3]

- An active tag relies on an internal battery for power. The battery is used to communicate to the interrogator, to power on-board circuitry, and to perform other functions. Active tags can communicate over greater distance than other types of tags, but they have a finite battery life and are generally larger and more expensive. Since these tags have internal power, they can respond to lower power signals than passive tags. [3]

- A semi-active tag is an active tag that remains dormant until it receives a signal from the interrogator to wake up. The tag can then use its battery to communicate with the interrogator. Like active tags, semi- active tags can communicate over a longer distance than passive tags. Their main advantage relative to active tags is that they have a longer battery life. The waking process, however, sometimes causes an unacceptable time delay when tags pass interrogators very quickly or when many tags need to be read within a very short period of time. [3]

- A semi-passive tag is a passive tag that uses a battery to power on-board circuitry, but not to produce return signals. When the battery is used to power a sensor, they are often called sensor tags. They typically are smaller and cheaper than active tags, but have greater functionality than passive tags because more power is available for other purposes. Some literature uses the terms "semi-passive" and "semi- active" interchangeably. [3]

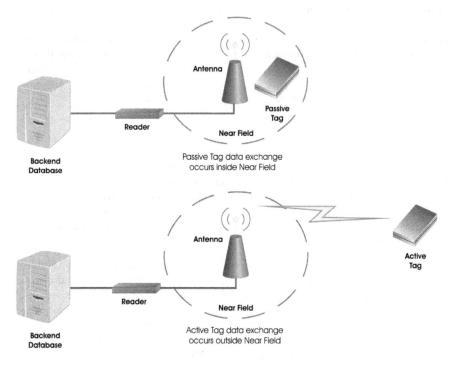

Figure 1. Passive and Active Tag Processes

Like bar codes in an earlier time, RFID is the next revolution in AIDC technology. Most of the advantages of RFID are derived from the reliance on radio frequencies rather than light (as is required in optical technology) to transmit information. This characteristic means that RFID communication can occur:

- Without optical line of sight, because radio waves can penetrate many opaque materials,

- At greater speeds, because many tags can be read quickly, whereas optical technology often requires time to manually reposition objects to make their bar codes visible, and

- Over greater distances, because many radio technologies can transmit and receive signals more effectively than optical technology under most operating conditions. [3]

2.1.1. Carrier frequencies

Today, there are four carrier frequencies implemented for RFID that are popular globally: 125 KHz, 13.56 MHz, UHF ranging from 866 to 950 MHz depending on national radio regulations, and microwave frequencies of 2.45 GHz and 5.8 GHz. There is also the frequency range 430-440 MHz, which is allocated to amateur radio usage around the world. The ISM band 433.05-434.790 MHz is located near the middle of the amateur radio band. The amateur radio band has emerged as an RFID channel in a number of applications. The frequency range has been called the 'optimal frequency for global use of Active RFID'. [1]

2.1.2. Functionality

- The primary function of a tag is to provide an identifier to an interrogator, but many types of tags support additional capabilities that are valuable for certain business processes. These include:

- Memory - memory enables data to be stored on tags and retrieved at a later time. Memory is either write once, read many (WORM) memory or re-writeable memory, which can be modified after initialization. Memory enables more flexibility in the design of RFID systems because RFID data transactions can occur without concurrent access to data stored in an enterprise subsystem. However, adding memory to a tag increases its cost and power requirements.

- Environmental sensors. The integration of environmental sensors with tags is an example of the benefit of local memory. The sensors can record temperature, humidity, vibration, or other phenomena to the tag's memory, which can later be retrieved by an interrogator. The integration of sensors significantly increases the cost and complexity of the tags. Moreover, while many tag operations can be powered using the electromagnetic energy from an interrogator, this approach is not workable for sensors, which must rely on battery power. Tags typically are only integrated with sensors for high-value, environmentally sensitive, or perishable objects worthy of the additional expense.

- Security functionality, such as password protection and cryptography. Tags with onboard memory are often coupled with security mechanisms to protect the data stored in that memory. For example, some tags support a lock command that, depending on its implementation, can prevent further modification of data in the tag's memory or can prevent access to data in the tag's memory. In some cases, the lock command is permanent and in other cases, an interrogator can "unlock" the memory. EPCglobal standards, International Organization for Standardization (ISO) standards, and many proprietary tag designs support this feature. Some RFID systems support advanced cryptographic algorithms that enable authentication mechanisms and data confidentiality features, although these functions are most commonly found on RFID-based contactless smart cards and not tags used for item management. Some tags offer tamper protection as a physical security feature.

- Privacy protection mechanisms. EPC tags support a feature called the kill command that permanently disables the ability of the tag to respond to subsequent commands. Unlike the lock command, the kill command is irreversible. The kill command also prevents access to a tag's identifier, in addition to any memory that may be on the tag. While the lock command provides security, the primary objective of the kill command is personal privacy. RFID tags could be used to track individuals that carry tagged items or wear tagged articles of clothing when the tags are no longer required for their intended use, such as to expedite checkout or inventory. The ability to disable a tag with the kill command provides a mechanism to prevent such tracking.[1]

2.2. RFID reader (Interrogator)

The second component in a basic RFID system is the interrogator or reader, which wirelessly communicate with a tag. Readers can have an integrated antenna, or the antenna can be separate. The antenna can be an integral part of the reader, or it can be a separate device. Handheld units are a combination reader/antenna, while larger systems usually separate the antennas from the reader.

The reader retrieves the information from the RFID tag. The reader may be self-contained and record the information internally; however, it may also be part of a localized system such as a POS cash register, a large Local Area Network (LAN), or a Wide Area Network (WAN).

There is also Middleware, software that controls the reader and the data coming from the tags and moves them to other database systems. It carries out basic functions, such as filtering, integration and control of the reader. [1]

RFID systems work, if the reader antenna transmits radio signals. These signals are captured tag, which corresponds to the corresponding radio signal (Figure 2).

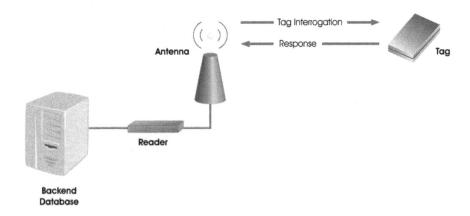

Figure 2. The interaction between the reader and RFID tag [2]

2.3. Security of RFID technology

Let's start with the first question: What are the security risks with RFID? The information inside [passive] RFID tags is vulnerable to alteration, corruption, and deletion due to low processing speed and low memory. In contrast, some high-end active RFID readers and tags tend to improve security through use of cryptography, challenge-response protocols, rotating passwords, and tamper detection technology. These devices have more processing power and more memory than their passive counterparts. They are more expensive and need a battery to give a boost to the processing power. The passive RFID devices do not need a battery. The tags wake up when they receive a signal from a reader.

Now let's go the second question: How can we categorize the attacks on RFID technology? The management can start with the four categories of the attacks that are unique to the RFID infrastructure: war-walking and lifting, counterfeiting, denial-of-service, and weak cryptography.

2.3.1. War-walking and war-lifting

War-driving, also known as the wireless LAN driving is a technique of using a Wi-Fi-based laptop or PDA to detect Wi-Fi wireless networks while driving in a vehicle, such as a small truck or an automobile. Legitimate war-drivers do not use services without proper authorization.

In the RFID technology arena, we add the wireless RFID driving to the description of war-driving. It is not necessary to have a LAN as an access point that a remote wireless device can pick up. A war-driver can use the device to pick up the information from unsecured tags affixed to an item, case, or pallet. What is more is that the war-driver could disable the RFID deactivation mechanisms when the items leave the retail stores.

In addition, the war-driver can read and get the information from the RFID tags of purchased goods that a passerby carries in a shopping bag. This can happen only if the tags are not properly deactivated when they leave a retail store or a warehouse.

War-walking is more bold than war-driving. War-walkers do not need a wireless device to find the RFID tags. With fake credentials or cards, they can bypass physical checks and find the system that uses RFID tags to monitor the movements of conference attendees.

Let's assume the cracker goes beyond finding the system. The cracker either runs away or removes the passive RFID tags from the objects, say, inside one case by sawing or etching the tags away. The cracker replaces them with the counterfeited tags, and reattaches the tag with original RFID data to the like objects in another case, all without being detected. This technique is known as lifting.

In another instance, a corporate spy walks around, scans the entire stock of a competing retail outlet, rewrites the tags of cheap products and replaces with better product labels and even hides products in a metal-lined tag and replaces with new tags on shelf. Passive tags do not work very well when they come into contact with a metallic surface.

Another privacy issue that has raised is what flashes up on a scanner as someone walks near the interrogator (especially the active interrogators that have a much wider scanning region than those of passive interrogators). The scanner could show:

- Clothing origins

- Contents of origins

- Contents of briefcase or handbag

- Which credit cards being carried

- Linkage to RFIDs that identify the user of passport in suit pocket

Make sure the RFID infrastructure is secured with physical security control mechanisms. If the company can afford it, it could use, for example, AXCESS's ActiveTag system, a single-system approach to automatic monitoring and tracking applications right from your desktop computer, including Asset Management, Personnel and Vehicle Access Control, Personnel Monitoring, Production and Process Control, and Inventory Tracking.

2.3.2. Counterfeiting

It is the semi-conductor companies who manufacture RFID tags. Unlike security firms, the semi-conductors have practically no experience in security. These companies are more interested in getting the customers to buy their products rather than in the discussion of product vulnerabilities and countermeasures. Another problem is the vendors who become too overconfident that their products will not be easy to break.

With a switched reader, you will be not able to read the tags. An adversary can defeat an encryption by switching readers after gaining physical access to the location that sends encrypted communications.

Now, how does an adversary make the switch? One possibility is to switch with a fake reader. Another possibility is to tamper with the original reader. It is so easy to do so with a portable handheld device, particularly the ones that can fit into the palm of most hands. The tampered or replaced reader can be modified to allow the adversary to control a legitimate reader nearby from a distance and write counterfeit serial numbers on the RFID tags. It also can be modified to automatically change the original RFID numbers stored in the reader's database and replace it with invalid numbers.

That is why it is important to secure custody for the reader even when a RFID handler is not using the device. It is also important for the organizations to ensure that a legitimate reader can reject an invalid RFID number counterfeited on the tag or in the reader's database.

You should determine what countermeasures you need to mitigate the risks of counterfeiting threats before RFID is fully implemented.

2.3.3. Denial of service

RFID radio signals area also very easy to block or jam. This can cause denial-of-service not only to the RFID tags but also at the data and network level.

Hackers and crackers can launch a denial-of-service attack by using electromagnetic fog to block RFID scanning and flooding a retail outlet with radio waves at the same frequencies as RFID scanners, thus causing chaos at check-outs. They also can hide a transmitter in a cat at a parking lot. This transmitter can block radio signals, causing an RFID-enabled store to close, and send a malicious virus to an EPC IS server containing the RFID data.

2.3.4. Weak cryptography

Although we expect the price for passive tags to drop below five cents per unit in a few years, we must acknowledge that these tags are computationally weak for the standard basic symmetric key cryptographic operations. Because more expensive RFID tags have more processing power and memory they can perform advanced cryptographic functions. Most low-cost tags are readable; many have limited writeable capability. This is because these tags are designed with basic functionality to keep the costs low.

Although we can get around this problem in a limited way via minimalist cryptography and Elliptic Curve Cryptography (ECC), they are more appropriate for other RFID devices, smart cards.

To overcome some of the confusing policies on when to use the kill command, the AUTO-ID Center and EPCglobal have proposed to put thef chip tags to sleep for a while rendering them inoperable temporarily and: then wake up these tags later on with a pair of sleep/wake commands.

As mentioned previously, the basic functionality of the low-cost RFID tags does not allow the basic cryptographic operations, due to limited processing power and little memory and size of the chip. To make it work, the tag must have memory of several megabytes and be rateable. The scheme for this cryptography is pseudonym throttling. It sores a short list of random identifiers of pseudonyms and goes into a cycle. Very little computation, if any, is involved, as contrasted to standard cryptography that requires quite a bit of computation and more complex circuitry.

The ECC is widely accepted for its efficient deployment of the public key mechanism. ECC is known for its compactness due to the novel way it uses arithmetic units to perform complex computations. It is much more compact then RSA, allowing the low-cost tags to be RFID-enabled. To get the ECC to work properly in RFID tags, we cannot overlook three important things: an adequate memory, the size of the area into which the ECC is installed, and the amount of power the tag can consume and emit signals to perform a simple computation. If the memory is too low, the ECC will not work. If the memory is adequate but the circuitry does not give enough power to consume, the ECC will not work. If the size of the area is too small regardless of memory size or the amount of power consumption, the ECC will not work. The memory, the area size, and power consumption, must be set properly in order for all three to get the computation to work properly.

2.3.5. Defence in depth

Let's assume light-weight cryptography for the RFID tag is well designed and is one of the protection mechanisms to defend the RFID infrastructure from attacks. In reality, 100 percent protection from cryptography is not possible. What is possible is the mitigation of risks to cryptographic attacks to an acceptable level. Another possibility is to let other protection mechanisms take over at the software/hardware level if one protection mechanism degrades or fails. They include firewalls, intrusion detection systems, scanners, RFID monitoring, failover servers, VPNs, and PKI.

As shown in Figure 3., these protections form the core of the Defense-in-Depth model of three rings. The middle ring focuses on access and audit controls. Access controls are best achieved with a WSSO for each user via SAML Auditing is accomplished with an examination of security practices and mechanisms within the organization.

Overlapping the core and middle rings are the operating systems that include both, for example, firewalls and access controls, such as Windows 2000 security, Windows 2003 Server Security, UNIX and Linux security, and Web security. Also included are the automated tools and devices to assess network vulnerabilities.

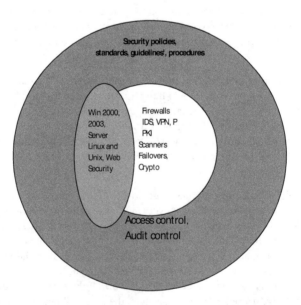

Figure 3. RFID Defence-In-Depth

The outer ring is a set of security policies including business continuity policy, risk assessment policy, password protection management policy, and server security policy.

Implementing the Defense-in-Depth is not as easy as it seems. Administrators must often choose from among a dizzying array of specialized hardware and software products to meet their organizations' need for network security.

To realize both best-of-breed application choice and full management integration, network administrators should consider an enterprise security solution built on an open architectural platform. With well-defined interfaces, this enables third-party security applications to plug in seamlessly with the overall security policy. In addition, an open architecture can leverage Application Programming Interfaces (APIs) to develop and deploy custom applications to meet specific network security needs.

2.4. RFID data collection tool-backend communication attack

Middleware and backend communication occur using JMS, SOAP, or HTTP. There are two types of attacks that can have an impact on the backend: MIM application layer attack and a TCP replay attack.

2.4.1. MIM Attack

A MIM attack occurs when someone monitors the system between you and the person you are communicating with. When computers communicate at low levels of the network layer, they may not be able to determine who they are exchanging data with. In MIM attacks, someone assumes a user's identity in order to read his or her messages. The attacker might be actively replying as you to keep the exchange going and to gain more information. MIM attacks are more likely when there is less physical control of the network (e.g., over the Internet or over a wireless connection).

2.4.2. Application layer attack

An application layer attack targets application servers by deliberately causing a fault in a server's operating system or applications, which results in the attacker gaining the ability to bypass normal access controls. The attacker takes advantage of the situation, gaining control of your application, system, or network, and can do any of the following:

- Read, add, delete, or modify your data or operating system,

- introduce a virus program that uses your computers and software applications to copy viruses throughout your network,

- introduce a sniffer program to analyze your network and gain information that can eventually be used to crash or corrupt your systems and network,

- abnormally terminate your data applications or operating systems,

- disable other security controls to enable future attacks.

- The best way to prevent MIM and application layer attacks is to use a secure way.

2.4.3. TCP replay attack

A replay attack is when a hacker uses a sniffer to grab packets off the wire. After the packets are captured, the hacker can extract information from the packets such as authentication information and passwords. Once the information is extracted, the captured data can be placed back on the network or replayed. Some level of authentication of the source of event generator can help stop TCP replay attacks.

2.4.4. Attacks on ONS

ONS is a service that, given an EPC, can return a list of network-accessible service endpoints pertaining to the EPC in question. ONS does not contain actual data regarding the EPC; it contains only the network address of services that contain the actual data. This information should not be stored on the tag itself; the distributed servers in the Internet should supply the information. ONS and EPC help locate the available data regarding the particular object.

2.4.4.1. Known threats to DNS/ONS

Since ONS is a subset of Domain Name Server (DNS), all the threats to the DNS also apply to ONS. There are several distinct classes of threats to the DNS, most of which are DNS-related instances of general problems; however, some are specific to peculiarities of the DNS protocol.

- *Packet Interception—Manipulating Internet Protocol (IP) packets carrying DNS information* Includes MIM attacks and eavesdropping on request, combined with spoofed responses that modify the "real" response back to the resolver. In any of these scenarios, the attacker can tell either party (usually the resolver) whatever it wants them to believe.

- *Query Prediction—Manipulating the Query/Answer Schemes of the User Datagram Protocol (UDP)/IP Protocol* These ID guessing attacks are mostly successful when the victim is in a known state.

- *Name Chaining or Cache Poisoning* Injecting manipulated information into DNS caches.

- *Betrayal by Trusted Server* Attackers controlling DNS servers in use.

- *Denial of Service (DOS)* DNS is vulnerable to DOS attacks. DNS servers are also at risk of being used as a DOS amplifier to attack third parties.

- *Authenticated Denial of Domain Names*

2.4.4.2. ONS and confidentiality

There may be cases where the Electronic Product Code (EPC) of an RFID tag is regarded as highly sensitive information. Even if the connections to EPCIS servers were secured using Secure Sockets Layer (SSL) /Transport Layer Security (TLS), the initial ONS look-up process was not authenticated or encrypted in the first place. The DNS-encoded main part of the EPC, which identifies the asset categories, will traverse every network between the middleware and a possible local DNS server in clear text and is susceptible to network taps placed by internet service providers (ISPs) and governmental organizations.

2.4.4.3. ONS and integrity

Integrity refers to the correctness and completeness of the returned information. An attacker controlling intermediate DNS servers or launching a successful MIM attack on the communication could forge the returned list of Uniform Resource Identifiers (URIs). If no sufficient authentication measures for the EPCIS are in place, the attacker could deliver forged information about this or related EPCs from a similar domain.

2.4.4.4. ONS and authorization

Authorization refers to protecting computer resources by only allowing the resources to be used by those that have been granted the authority. Without authorization, a remote attacker can do a brute-force attack to query the corresponding EPCIS servers until a match is found. In case the complete serial number is not known, the class identifier of the EPC may be enough to determine the kind of object it belongs to. If using the EPCglobal network becomes ubiquitous and widespread, the attacker could add fake serial numbers to the captured, incomplete EPC and query the corresponding EPCIS servers to find a match. This can be used to identify assets of an entity, be it an individual, a household, a company, or any other organization. If you wore a rare item or a rare combination of items, tracking you could be accomplished just by using the object classes.

2.4.4.5. ONS and authentication

Authentication refers to identifying the remote user and ensuring that he or she is who they say they are.

2.4.4.6. Mitigation attempts

• *Limit Usage* Use the ONS only in intranet and disallowing any external queries.

• *VPN or SSL Tunneling* With data traveling between the remote sites, it needs to be exchanged over an encrypted channel like VPN or SSL Tunneling.

• *DNS Security Extensions (DNSSEC)* ensure the authenticity and integrity of DNS. This can be done using Transaction Signatures (TSIG) or asymmetric cryptography with Rivest, Shamir, & Adleman (RSA) and digital signature algorithms (DSAs).

2.5. Risk and vulnerability assessment

The assessment of risks and vulnerabilities go hand in hand. To begin evaluating your system, you need to ask questions regarding the assessment and tolerance of the risks: what types of information are you talking about at any given point in the system and what form is it in? How much of that information can potentially be lost? Once these risks are evaluated, you can begin to plan how to secure it. A good way to evaluate the risk is to ask five classic investigative questions: "who?", "what?", "when?", "where?" and "how?"

- **Who** is going to conduct the attack or benefit from it? Will it be a competitor or an unknown group of criminals?

- **What** do they hope to gain from the attack? Are they trying to steal a competitor's trade secret? If it is a criminal enterprise, are they seeking customers' credit card numbers?

- **When** will the attack happen? When a business is open 24 hours a day, 7 days a week, it is easy to forget that attacks can occur when you are not there.

- **Where** will it take place? Will the attack occur at your company's headquarters or at an outlying satellite operation? Is the communications link provided by a third party vulnerable?

- **How** will they attack? If they attack the readers via an RF vulnerability, you need to limit how far the RF waves travel from the reader. If the attacker is going after a known vulnerability in the encryption used in the tag reader communications, you have to change the encryption type, and, therefore, also change all of the tags.

2.5.1. Type of RFID risks

RFID technology enables an organization to significantly change its business processes to:

- Increase its efficiency, which results in lower costs.

- Increase its effectiveness, which improves mission performance and makes the implementing organization more resilient and better able to assign accountability, and

- Respond to customer requirements to use RFID technology to support supply chains and other applications.[16]

This section reviews the major high-level business risks associated with RFID systems so that organizations planning or operating these systems can better identify, characterize, and manage the risk in their environments. The risks are as follows:

Business process risk - direct attacks on RFID system components potentially could undermine the business processes the RFID system was designed to enable. For example, a warehouse that relies on RFID to automatically track items removed from its inventory may not be able to detect theft if the RFID system fails.

Business intelligence risk - an adversary or competitor potentially could gain unauthorized access to RFID-generated information and use it to harm the interests of the organization implementing the RFID system. For example, an adversary might use an interrogator to determine whether a shipping container holds expensive electronic equipment, and then target the container for theft when it gets a positive reading.

Privacy risk - the misuse of RFID technology could violate personal privacy when the RFID application calls for personally identifiable information to be on the tag or associated with the tag. For example, if a person carries products that contain RFID tags, those tags may be surreptitiously read by an adversary. This could reveal that person's personal preferences such as where they shop, or what brands they buy, or it might allow them to track that person's location at various points in time.[16]

Externality risk - RFID technology potentially could represent a threat to non-RFID networked or collocated systems, assets, and people. For example, an adversary could gain unauthorized access to computers on an enterprise network through Internet Protocol (IP) enabled interrogators if the interrogators are not designed and configured properly. Multiple RFID interrogators operating in a confined space may cause hazards of electromagnetic radiation to fuel, ordinance or people in the vicinity.

2.5.2. Risks in supply chain management and tracking applications

Tracking applications are used to identify the location of an item, or more accurately, the location of the last interrogator that detected the presence of the tag associated with the item. An example of an intentional attack on an RFID business process is cloning, which occurs when an adversary reads information from a legitimate RFID tag and then programs another tag or device to emulate the behavior of the legitimate tag. Another attack on an RFID business process would be removing a tag from the item it is intended to identify and attaching it to another unrelated item.

Supply chain management involves the monitoring and control of products from manufacture to distribution to retail sale. Supply chain management typically bundles several application types, including asset management, tracking, process control, and payment systems. Supply chain systems record information about products at every stage in the supply chain. Ideally, tags are affixed to products during the manufacturing process or soon afterward. As a product moves through the supply chain, to the customer, and to post-sale service, the tag's identifier can be used by all supply chain participants to refer to a specific item.

In addition, supply chain systems that use active tags can track larger objects such as cargo containers. Tags on these containers can store a manifest of the items shipped in each container. This manifest can be automatically updated when items are removed from the container. Potential problems are not just limited to the RF subsystem. If the network supporting the RFID system is down, then the RFID system is likely down as well. In supply chain applications, network failures at any point in the chain have the potential to impact the business processes of any subsequent link in the chain. For example, if a supplier is unable to write manifest data to a tag, then the recipient cannot use that data in its operations even if its RFID interrogators and network infrastructure are fully functional. Servers hosting RFID middleware, databases, analytic systems, and authentication services are all points of failure.

Any efforts to assess business process risk need to be comprehensive, because such a wide variety of potential threats exist. All of these threats have the potential to undermine the supported business process and therefore the mission of the implementing organization.[3]

3. RFID in procedural conditions of logistic operators

Supply chain can be defined as the parts that are involved, directly or indirectly, in fulfilling a customer request (Chopra and Peter 2007). By this definition, it can be seen that a supply chain consists of manufacturers, warehouses, retailers, transporters, and customers. The purpose of a supply chain is to maximize the value generated for the customer; namely, maximizing the difference between the final product worth and the total expended by the supply chain to provide the product to the customer.

In order to succeed, the supply chain must be conducted to minimize the costs incurred. Supply chain management (SCM) is responsible for optimizing the flows within its operational stages which include raw materials, manufacturing, distribution, and transportation in order to minimize the total cost of the supply chain. SCM is a unification of a series of concepts about integrated business planning that can be joined together by the advances in information technology (IT) (Shapiro 2007), yet many companies have not completely taken advantage of this process.

In today's world, the competition between companies, more demanding customers, and reduced margins make the scenario more difficult for companies to succeed, to this context, SCM is an important practice for companies that want not only to keep in business but also have their results optimized and meet the clients' expectations.

Responsiveness in the supply chain has gained importance and it is a trend that apparently will dictate future decisions regarding supply chain design. According to Kovack, Langley, and Rinehart (1995), the themes that will have influence on logistics on the near future are:

- Strong corporate leadership will enhance logistics value through focusing on efficiency, effectiveness, and differentiation.

- Value realization requires marketing of logistics capabilities within the company and to external customers.

- Emphasis on the "scientific" aspect of logistics management in order to enhance the "art" of creating customer satisfaction. Enhancing logistics value through integrating product, information, and cash flows for decision-making linking external and internal processes. Logistics value enhanced by ownership of responsibility internally and externally to the firm.

- Focus of successful companies is to create internal value for their organizations and external value for their suppliers and customers.

From these themes, it can be seen that SCM plays and will continue to play an active role in successful companies' routines. In order to achieve better results in the supply chain and better responsiveness to customers' necessities, new techniques such as real-time inventory and dynamic supply chain need to be developed.

3.1. Transportation in logistics and SCM

As a supply chain driver, transportation has a large impact on customer responsiveness and operational efficiency. Faster transportation allows a supply chain to be more responsive but reduces its efficiency. The type of transportation a company uses also affects the inventory and facility locations in the supply chain. The role of transportation in a company's competitive strategy is determined by the target customers. Customers who demand a high level of responsiveness, and are willing to pay for the responsiveness, allow a company to use transportation responsively. Conversely, if the customer base is price sensitive, then the company can use transportation to lower the cost of the product at the expense of responsiveness. Because a company may use transportation to increase responsiveness or efficiency, the optimal decision for the company means finding the right balance between the two.

The transportation design is the collection of transportation modes, locations, and routes for shipping. Decisions are made on whether transportation will go from a supply source directly to the customer or through intermediate consolidation points. Design decisions also include whether multiple supply or demand points will be included in a single run or not. Also, companies must decide on the set of transportation modes that will be used.

The mode of transportation describes how product is moved from one location in the supply chain network to another. Companies can choose between air, truck, rail, sea, and pipeline as modes of transport for products. Each mode has different characteristics with respect to the speed, size of shipments (parcels, cases, pallet, full trucks, railcar, and containers), cost of shipping, and flexibility that lead companies to choose one particular mode over the others. Typical measurement for transportation operations includes the following metrics:

• Average inbound transportation cost, or the cost of bringing product into a facility as a percentage of sales or cost of goods sold (COGS). Cost can be measured per unit brought in but is typically included in COGS. It is useful to separate this cost by supplier.

• Average incoming shipment size measures the average number of units or dollars in each incoming shipment at a facility.

• Average inbound transportation cost per shipment measures the average transportation cost of each incoming delivery. Along with the incoming shipment size, the metric identifies opportunities for greater economies of scale in inbound transportation.

• Average outbound transportation cost measures the cost of sending product out of a facility to the customer. Cost should be measured per unit shipped, oftentimes measured as a percentage of sales. It is useful to separate this metric by customer.

• Average outbound shipment size measures the average number of units or dollars on each outbound shipment at a facility.

• Average outbound transportation cost per shipment measures the average transportation cost of each outgoing delivery.

- Fraction transported by mode measures the fraction of transportation (in units or dollars) using each mode of transportation. This metric can be used to estimate whether certain modes are overused or underutilized.

- The fundamental trade-off for transportation is between the cost of transporting a given product (efficiency) and the speed with which that product is transported (responsiveness). Using fast modes of transport raises responsiveness and transportation cost but lowers the inventory holding cost.

3.2. Information technology and SCM

It is no surprise that IT played a big role in enabling many processes and ideas in Supply Chain Management (SCM) that seemed impossible in earlier years. The first advance was the decreasing of inventory levels by managers abandoning rules of thumb and adopting the setting of inventories based on service level desired and historical demand (Shapiro 2007). IT allowed the analysis of a great quantity of units and the process of recalculating the inventory level as the demand changed. This ability to analyze inventory needs made the companies more agile while decreasing inventory levels and increasing service levels.

Another important fact that gave a great contribution to SCM was the electronic interchange (EDI). This technology allows the direct data interchange between companies using computers. EDI changed the relationship between the company and customers, with its suppliers, and also with the employees. The ability of trading data almost instantly across the supply chain gave companies the ability to manipulate more up-to-date information in a shorter period of time. This reduced the need for printing and transporting papers, enabled just-in-time practices, and helped to restructure logistics supply chain relationships. Together with EDI we can also mention the importance of the Internet in global business (Johnson et al. 1999).

Artificial intelligence systems are responsible for many advances achieved by society and by SCM as well. Computers can be programmed to execute routine functions and according to the rules imposed to the computer it can be capable of behaving an intelligent system that can execute complex activities in reduced time. This brought to logistics a much larger capacity of processing information and executing tasks. Many activities can operate without human interference and this converges to a more responsive and accurate supply chain (Johnson et al. 1999).

Some technologies, discussed later in this chapter, can be used to make real-time adjustments to the supply chain. Those adjustments could be due to many events such at manpower shortages or equipment breakdowns. For example, if a problem occurs with a truck or the road conditions change due to weather, the system, supplied with this updated information, should be able to make the necessary corrections to the transportation routes of other trucks to compensate for the truck failure.

This system would be very useful for natural disasters such as Hurricane Katrina. With real-time information, the system would reallocate transportation and production. This kind of modeling would reduce the response time for such events from months or weeks to days or even hours. This system can also be expanded to urban transportation within a city or long distances between two cities.

3.2.1. Real-time technologies

Radio frequency identification (RFID) and global positioning systems (GPS) are emerging technologies that will allow for real-time data collection to assist with decision support in SCM. RFID has a wide variety of applications. Some examples of RFID uses are library checkout stations, automatic car toll tags, animal identification tags, and inventory systems. Real-time data collected using RFID allows a supply chain to synchronize reorder points and other data. Real-time information can also be used to design and operate logistical systems on a real-time basis. GPS is currently used solely as a means to locate equipment and derive navigation directions.

An RFID system consists of a reader, tags, and an air interface. The reader, also known as an interrogator, sends out a signal through an antenna. This signal is usually in the form of an electromagnetic wave, so a direct line of sight is not needed to read the information on the tag. This is a major advantage of RFID. The signal is received by the tag and a response signal is sent back to the reader. This response signal contains a unique identifier associated with a tag. The response signal can be powered in two ways corresponding to the type of tag. Passive tags utilize the energy of the original signal to send a response signal back to the reader. Passive tags have a limited amount of energy to power the response signal. Therefore, the amount of information transmitted by a passive tag is fairly small, quite similar to the information carried in a bar code. Active and semi-active tags use energy from an attached battery to power the response signal. The use of the embedded battery allows the response signal to contain more information and travel farther. The reader receives the response signal, decodes it, and sends that information to a database. Often the information in the response signal is connected to additional information in the database.

RFID technology can be used throughout the supply chain in order to promote visibility. This visibility helps coordinate actions between entities in the supply chain. Figure 4 shows the relationships within the supply chain that can be affected by the implementation of the RFID technologies. An example of RFID implementation is the use of active tags for detecting tampering and monitoring security of maritime containers. Those types of tags also have the tracking advantages of RFID and can be used to improve operations management. Those tags can be seen in Figure 5.

GPS systems consist of a series of receivers and satellites that orbit the Earth-GPS works by calculating the distances from a receiver to a number of satellites. With each distance between a receiver and satellite, the number of possible locations is narrowed down until there is only one possible location. A receiver must calculate its distance from at least three satellites to determine a location on the surface of the Earth. However, four satellites are usually used to increase the location accuracy (Dommety and Jain 1996). This process of location would be controlled by the positioning module of GPS system. An average GPS positioning and navigation system would also have the following modules:

- **Digital map database,**
- **Map matching,**
- **Route planning and guidance,**
- **Human-machine interface,**
- **Wirelesscommunication.**

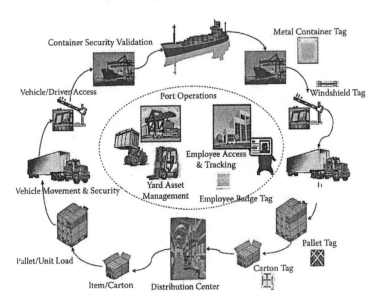

Figure 4. Integrated supply chain with RFID (Source: SAVI Technology)

There are three positioning technologies that can be used: radio wave-based positioning, dead-reckoning, and signpost. The use of GPS for navigation can have direct and indirect impacts on intelligent transportation systems. GPS navigation systems can provide information about local surroundings. Also, emergency personnel can be provided with a precise location for situations, thus reducing response times. Asset tracking is one of the most popular uses of GPS. One of the limitations of GPS is that receivers cannot communicate with satellites when indoors (Feng and Law, 2002).

RFID and GPS are radio wave-based technologies that are currently used by many organizations. RFID is primarily used in inventory and material handling processes. Tags are placed on items. When these items pass by checkpoints where readers are located, the tag is read and the appropriate action can be taken. Real-time inventory can be kept by monitoring tag reads at strategic points like loading docks. RFID can also be useful in material handling. Items on a conveyor can be diverted at the appropriate times based on the information received from the RFID tag. GPS is primarily use to track assets such as vehicles and other expensive equipment. For example, if a truck breaks down, it is possible to locate the truck and get the shipment moving again in a fraction of the time it would take with a GPS receiver.

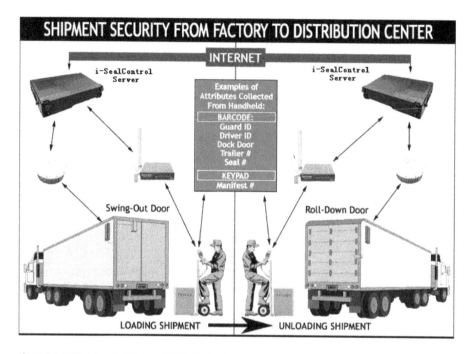

Figure 5. RFID Container Seal (Source : SAVI Technology)

Summary of RFID and Information Enablers

This section provides understanding of key technologies and how all the technologies differ and how they can be integrated to work for operational effectiveness. This will allow warehouse management system algorithms such as "bucket brigade" calculations, picking route optimization, and other effective system updates that will improve operations. Further insights into safety stock minimization, customer order optimization, and pick/stock labor minimization will be affected and discussed later.

3.2.2. RFID Provides timely visibility in logistics

RFID supports information in the supply chain by enabling visibility. The concept of visibility describes the ability of anyone, including customers, to have access to inventory, orders, raw materials, and delivery points at any time. Visibility is currently [provided by a mixture of automatic identification, or auto-ID, technologies such as bar codes, smart labels, ISBN, and UPC codes, along with others. The opportunity for RFID is that its non-line-of-sight scanning, the integration of the aforementioned auto-ID identifiers into RFID nomenclature, and the push for standardized technology protocols will provide large supply chain savings.

The real-time nature of RFID is considered a benefit and currently a challenge. The benefit is that you have the latest information to make the best decisions; the drawback is that the amount of data currently presents a data storage problem for operational systems.

Better visibility provides reduced inventory, labor and assets management using inventory policies, scheduling, and decision support system information. This is exemplified by the fact that:

- RFID supports reduced inventory costs with more effective labor policies

- RFID supports labor reduction with more effective scheduling

- RFID supports the reduction of expensive assets such as facilities, trucks, containers, and railroad time because of more accurate information in decision support systems. The ability for RFID to provide timely information and visibility into the supply chain are based on three components of RFID technologies. They are

- **Automatic data capture,**

- **Real-time information**

- **Real-time location system.**

The RFID enabling technologies diagram shown in Figure 6 represents these components as interconnecting orbits.

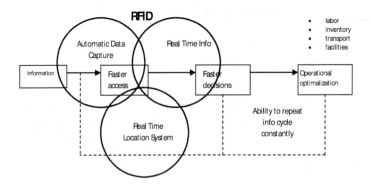

Figure 6. RFID-enabling technology

The figure also shows how RFID supports timely information in the supply chain by enabling information to be accessed faster. This implies that faster decisions can be made, which produces operational optimization that can be effectively repeated. In the figure, one of the boxes represents the RFID information flow. The ability to allow resident information collected automatically in real-time leads to faster, more effective decisions is where RFID shows future promise. Business costs are reduced as operations become more productive by reducing labor, transportation, and facility cost of moving inventory in the supply chain and postal services.

Many organizations see that the benefit of using RFID is that they can effectively manipulate inventory. Inventory exists in the supply chain because of the variance between supply and demand. This variance is necessary for manufacturers where it is economical to manufacture in large lot quantities and then store for future sales. The variance is also present in retail stores where inventory is held for future customer demand. Oftentimes businesses suggest that inventory is a marketing vehicle creating demand by passing customers. The main role inventory plays is to satisfy customer demand by having product available when the customers want it. Another significant role that inventory plays is reducing cost by exploiting economies of scale that may exist during production and distribution. Given that economy of scale is believed to have such a large impact on inventory, we will present some relevant information regarding inventory in the supply chain.

RFID is essentially in the same position occupied by mobility and wireless technology a few years ago. It is poised to spark a global revolution—in supply chain visibility and management. Using RFID in pivotal points in the supply chain can help enable a vision of having goods available to customers at the right place and at the right time. RFID technology is an enabler of this vision aiding the synchronization between physical and information flow of goods across the supply chain from Manufacturer to Retail Outlet, represented on figure 7. [1]

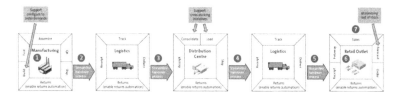

Figure 7. Supply chain containing RFID technology

Manufacturing

As goods travel down the production line, RFID tags are physically applied and a unique ID is written and then validated for quality assurance purposes. The unique ID is automatically associated to the product/order details to facilitate further tracking and exception management.

During the pallet build process; goods (e.g cases) are automatically identified to aid with customer order configurations. Finally, pallets are identified and tracked as they are delivered to the staging area ready for shipment.

Manufacturer – Logistics pickup

As the logistics vehicle arrives at the loading dock, the RFID reader positioned at the loading dock communicates with the unique RFID tag to confirm that the logistics vehicle is authorised to pickup goods. Upon approval, pallets leaving the loading dock communicate with the RFID reader to alert B2B systems (ASN) and ERP systems to initiate electronic transactions, proof of pickup and potentially shipment invoicing.

Logistics delivery – Distribution centre (dc)

As the logistics vehicle arrives at the Distribution Centre, the RFID reader and middleware initiates an event that captures the unique ID from the RFID tag, triggering the arrival of the manifest to initiate automatic routing of the goods to the next logistics vehicle (load consolidation).

Distribution centre – Logistics delivery

As pallets are loaded onto the logistics vehicle the RFID reader positioned above the loading dock communicates with the RFID tags. The RFID tags broadcast their unique ID to the reader and via the RFID middleware transfer information to ERP systems indicating that the manifest is loaded.

Logistics delivery – Retail outlet

As the shipments of goods arrive at the receiving dock (again being detected by RFID readers), Retail ERP systems are updated to manage inventory levels (automatically, accurately and at low cost) and initiate B2B messages to Suppliers to commence invoicing.

Retail outlet – Customer

As items are removed at shelf level, the RFID reader can automatically detect the event and via the RFID middleware, initiate additional product supply requests. With such a system in place, the need to maintain costly volumes in remote warehouses is almost eliminated. At this point of the process, the customer is initiating direct demand generation on the supply chain management process.

Customer

Rather than wait in line for a cashier, the customer simply walks out the door with the purchase. A reader built into the door recognises the items in the cart by unique ID's. A swipe of the debit or credit card and the customer is on their way.

3.2.2.1. Future technologies

Current applications of RFID and GPS systems have allowed for more effective tracking of inventory and assets. These technologies can be used in conjunction, but the data has to be captured and written to a database to be correlated to other tags or receivers. If these technologies can be combined to produce hybrid systems, greater gains can be achieved. One focus of research is the nesting of GPS receivers and various RFID tag types. If tags and receivers were able to communicate with one another, even more accurate real-time data collection could be achieved during transportation. This would also reduce equipment costs because fewer readers would be required. The nesting would follow the form in Figure 8.

If these technologies can be nested, it will allow the information in a bar code or a passive RFID tag to be collected by an active tag. This information could then be combined with the information contained within the active tag and transferred to a GPS receiver. The GPS receiver could then send not only its location but all of the information about the cargo being shipped (Reade and Lindsay 2003). A possible application of this nested technology approach would be in the railroad industry. Currently, there are two passive RFID tags attached to the sides of all railcars in the United States. In addition, most railroads use GPS receivers to track locomotives. If nesting became possible, implementation would be easy. Active tags could be used to capture the information correlated to the cargo in all of the railcars and transmit it to the GPS receiver and thus to the inventory databases.

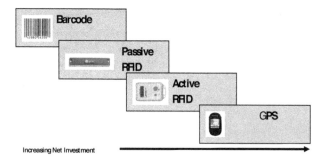

Figure 8. Nesting diagram

In addition to nesting technologies, more advanced tags can be developed to allow more detailed data collection. Tags that utilize sensors to capture and write data to the tag are being developed. Some tags have been developed but are still very unreliable. These sensor tags could be used to monitor physical parameters, like temperature and humidity, as well as security parameters. The main problem faced by these passive sensor tags is the limited power supply. The sensor cannot use any energy while outside the range of the reader. Also, the amount of energy available while in read range is very small. This limits possible measurement techniques (Want 2004). With these sensor tags, perishable goods could be monitored to guard against possible safety issues. This could include salmonella outbreaks caused by frozen chicken reaching too-high temperatures for too long and medications being held at temperatures that reduce potency.

4. AIDC and mobile technologies in postal sector

This part primarily deals with identification of postal items and transport units in logistic chain of postal operators. Nowadays, the identification is carried through barcodes and optical character recognition. In this article we would like to specify, how can be transport units identified in the transmission process by RFID technology. In the carriage of postal items is necessary to decide what type of transport is used for that purpose, what the flows of items are and what their intensity is.

The part described scheme of the transport process, including planned technology and there is also simulated a real postal process in conditions close to operational.

An unavoidable part of today is a dynamic development in the field of mobile technologies, their everyday use and application of the processes, which largely supports the level of quality of postal services and thereby strengthening the market positions of individual postal operators. This area is even more pertinent that in all countries of the European Union since 1 January 2013 approved the postal market and postal services. In this respect, it is necessary to include postal processes embarked on new technologies to ensure the competitiveness of the national postal operator and alternative providers.

RFID technology has been selected by an international post corporation (IPC) to test deliverability (transit time) of items in 55 countries of the world (Slovak republic including). The requirement of transit time is defined by Universal Postal Services and applicable also for Slovak Post. Despite the RFID technology is being known and being improved for a long time, it is essential to define the standards and security requirements.

Besides efficiency, consolidation and globalization within the European Union, interoperability is one of key elements. It is the ability of information and communication systems (including the supported processes) to exchange data, share information and knowledge, which leads to standardization.

4.1. Methods and aims

For understanding of issue is should be analyze terms used. The availability of RFID components, GPS devices and possibility of using satellite navigation there is possible to create a relative effective infrastructure for improving management of transport process by post.

There is true, that personal correspondence is on the wane, the main reason is development of information technology especially Internet, but large part of using a postal services have a companies and therefore the services will remain an indispensable part of society.

4.2. Structure

When we focus on these connections, external influences on postal sector and potential current technologies there is important to analyze possibilities of automation individual processes, improve a transportation operating activities and ensure continuity in fulfilling the goals. These aims lead to satisfying of customers in area of provide post services at phase in the delivery of mail.

The aim of this part is refer on possible improve in this area. The most important term of category, which will use in individual chapters are: mobile technology, definition of means transport.

4.2.1. Mobile technology

The classification of wireless technologies based on the distance or reach of the broadcast signal provides insight on their potential use. A condition of transport a date in broadcast systems and networks is communication without physical contact.One of the possible division of this system is on range of coverage:

- **Global system** – These systems coverage of territorial area. There we can speak on worldwide operating systems, which aren't dependent on a concrete application and their communication is carried through different protocol. (for example: Satellite communication systems, GPS)

- **Metropolitan systems** – These systems operate on lower geographic area. They usually operate at state level. (For example: The system based on wireless technology, Wi-Fi)

- **Local systems** - These systems operate at a distance, which include a several cm up to several hundred meters (For example: Bluetooth, RFID)

4.3. Postal transport network

The postal transport is most important part of process from submission of mail after its delivery to addresses with consistent set of quality standards for different types of mail. These standards are also based on the postal license and the requirement for quality by the universal postal services.

The postal transport network includes postal courses and infrastructure. While constructions of postal transport network are use a different systems and tools. The postal transport network is divided into three basic levels:

- **district transport network (OPS)** - this network connecting establishment with other facility of processing center area.

- **regional transport network this(RPS)** – this network connecting the main processing centers with district processing centers of own district.

- **main transport network (HPS)** – this network connecting the main processing centers, the main processing center with the district processing centers from different district of HSS. This network includes transport conclusion in international relations.

In the carriage of postal items is necessary to decide what type of transport used for that purpose, what are the flows of items and what is their intensity. Way to connect and type of vehicle depends on the following factors:

- density and organization of the postal network,

- flows of different types of postal items and their size,

- the carrying capacity of vehicles used,

- transport time of each species of postal items,

- safety and effectiveness of postal traffic.

Processing of items is implemented in the workplace of the Slovak Post:

- **HSS - main processing center** - the facility providing treatment and quest items posting its area of perimeter, mail items addressed to your district and in transit in its dealings with OSS circuit, in contact with other HSS and OU,

- **OSS - regional processing center** - post office responsible for preparing and quest items posted at post offices in his own constituency and in transit in contact with your postal district and interacted with the HSS, the facility responsible for receiving, processing and quest items express postal services,

- **selected post** - post office responsible for preparing and quest items selected species within a specified range (usually as OSS),

- **Exchange post** - processing the shipment and ensure shipments to post offices exchange foreign postal administrations,

Regional hub as department of express service - establishment is responsible for receiving, processing and quest items express postal service.

4.4. Characteristics of transport units and processes

Characteristics of transport units - Slovak Post, a. s. used in the transport process shipments following shipping units: containers, leaf containers and postal bags. Containers are used in the transport process at HSS and OSS, on the local network using only postal letter case and postal bags.

Basic flow of transport processes are show on next figure including use a basic mobil technology in relevant stages.

The postal courses represent connection, which is set by transportation route with time data movement of vehicles used for carrying of postal mails. The postal courses are divided by the following criteria:

- **The rail transport** – used on carrying of postal mails through rail network. The conclusions are transport in wagon, which owned SP, a.s.

- **The road transport** – used road infrastructure for carrying conclusions. The conclusions are transport by vehicle, which own of SP, a. s.

- **The fly transport** – this type of transport is most advantageous for fast speed and overcoming large distance. The SP, a.s. used this type of transport on agreements with individual airlines. It only use for international postal mail transport.

The greatest part of transport postal mail is ensured by the road transport between main transport network (HPS), regional transport network (RPS) and district transport network (OPS).

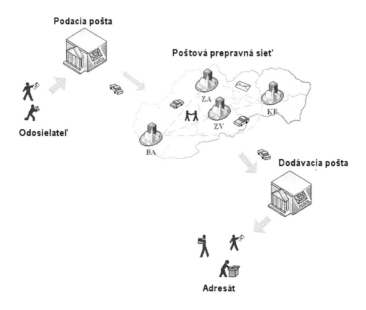

Figure 9. Simplified diagram of movement of the consignment of transmission network in Slovak

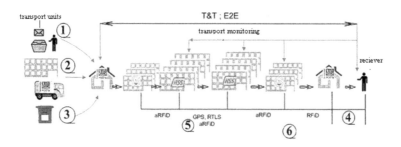

Figure 10. Scheme of the transport process, including planned technology

1. sender pass the post office at the counter

2. collecting expedition posting or accumulating courses,

3. pass through postal box,.

4. mobile technology - monitoring the transport process,

5. possibilities for optimizing routes for mail delivery

6. communication with the addressee.

4.4.1. Transport units

The Slovak Post used the following transporting units in the transport process:

- container

- letter boxes

- bags

The postal operator has four types of containers for transport of letters and bags:

- platform truck – made by aluminum profiles connecting by PVC parts. This container is equipped by securing straps,

- stable structure track with rear wall and two side panels with wire grid 100x100mm,

- truck shipments on a very stable structure, floor frame and rugged steel profile galvanized steel thickness 1mm,

- folding platform truck made of steel profiles welded together by fasteners.

The containers are used in the transport processing at HSS and OSS. In the local postal network used only containers and bags.

4.5. Design applications

It is obvious that these systems are in a lot of cases combined and interrelated. In this design is emphasis on technology, which their using isn't common. There is mean GPS, Wi-Fi, GSM and more. On the figure, there are plans with this technology. Some of these technologies the postal operators used now and this is reason, why was this design focused on mail monitoring in transport processes on passive RFID technology.

For possible future use of the possibilities currently offered by some mobile technologies, we have tried outline Figure 11 scheme of the transport process, including the applicable technologies selected and purpose of their use:

aRFID such as active RFID technology – application within HSS and OSS use on monitoring containers a transporting units, optimization process and better evaluation quality of postal services,

- **pRFID** such as passive RFID technology – application between post office and sender/ addresses of postal mails. There is a lot of option of using,

- **BC** – bar code – barcode using by SP, a, s. at present,

- **RTLS** – monitoring mails, which are important or contain perishable content,

- **GPS** – route monitoring, possible specifying of delivery place for some type of mails,

- **GSM** – communication through mobile phone, information about mails, possible locate a place for delivery mails, possible pay for service through mobile phone

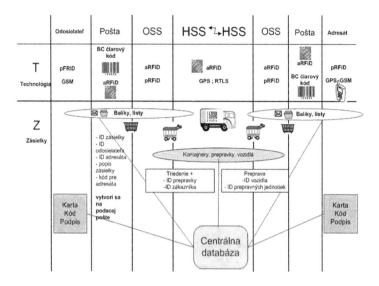

Figure 11. Scheme of the transport process, including the anticipated technologies

4.6. Types of RFID application in conditions of postal processes

The AMQM[1TM] Platform provides postal operators with a complete overview and effective traceability of mail volumes, parcels, mail bags, mail items, trucks, roll-containers and letter trays across the entire logistics chain. One key feature is automatic consignment system that associates the mail items to the containers carrying these items and to the trucks transporting these.

This solution can be based on various technologies such as: RFID, disposable RFID labels and bar codes, as well as combinations thereof. It also enable objective documentation of times of arrival and departure of vehicles, which postal containers are loaded/unloaded, vehicle load space management, real-time information on types of mail, quantities, times of arrival, delays or changes in transport times etc. With regard to postal operational systems, the following conditions must be taken into account:

• Rough industrial environments.

• Large volumes of goods and mail.

• Short time available for processing.

• High labor costs in connection with the daily operations.

• Large potentials in automation and streamlining of manual processes.

---

1 AMQM – Automatic Mail Quality Measurement

4.6.1. RFID-based vehicle management

Tracking vehicles and trailers throughout the entire transport logistics chain provides considerable benefits to all parties involved, e.g. management, users and customers. The Vehicle and Trailer Tracking System is an advanced and effective IT system for monitoring and managing precise arrivals and departures of vehicles at specific points in the logistics chain.

The system is built on the experience and know-how acquired from supplying the world's largest and most widespread RFID network stretching across about 60 countries.

Implementing this system offers unique values. Examples of benefits:

• Fully automatic registration of vehicles - i.e. no manual work involved.

• Improved yard and vehicle management.

• Precise and objective record of exchange of goods between parties.

• Early warning on delays in transport to all parties.

• Precise feedback to transport planning systems.

• Improved vehicle maintenance routines.

• Cost savings in centers with real-time information available.

4.6.2. Roll cage tracking and managing

One of the main issues being addressed by the roll container tracking and managing project is need to take control of and better manage transportation assets. Another primary project requirement is to ensure that the required containers will be always available at the customers' premises and within postal operator facilities. This should overcome the tendency for planned or unplanned hoarding of roll containers that causes shortages elsewhere, especially at peak times.

Additionally, the lack of visibility of roll container whereabouts led to unnecessary loss since it was not possible to identify where the roll containers disappeared and hence forced expensive purchase of new roll containers to meet the customer service level agreements. System of the monitoring and managing roll cages includes tag (active or passive, it depends of application), that is placed on a side or on the bottom of the container (Figure 12), it also includes a handheld terminal solution for consignment of roll container, product and destination enabling load control on all roll containers (Figure 13).

The result is avoiding miss-sending and has real-time t volume forecasting into all facilities in the network providing efficient and on time production and distribution. Also a handheld terminal solution designed for track and trace of all individual parcels is a part of the solution providing key customers with shipment visibility throughout the whole logistic network. Miss-shipments are prevented by load-control.

Figure 12. RFID tag placed on the container

Figure 13. Handheld terminal

When a roll container is ready for dispatch, the roll container is scanned for destination and product type. If the roll container is lead through a gate not matching the destination, an alert will immediately help correct the mistake. Solution must include Asset Management software platform enabling full, real-time transparency of the location of each roll container and can be also used to track specific mail and parcel transports. [6]. Implementing this system offers unique values. Examples of benefits:

- Improves availability and load balance throughout the logistics chain.

- Prevents hoarding of roll-containers.

- Minimizes losses.

- Helps to improve supply chain efficiency.

- Provides the ability to monitor the transported delivery time of goods.

- Helps to improve service and maintenance.

4.6.3. Letter tray tracking

Tracking and tracing letter trays throughout the entire postal logistics chain provides benefits to postal customers, employees and management. The trays are automatically registered in the postal logistics by means of RFID technology. Each letter tray has a tag that communicates and transmits information to the reader in Real-time load control (Figure 14). Now it is possible for the postal operators to reuse the same RFID network to track & trace postal letter trays. This new opportunity is a fast pay-back investment with many unique advantages to postal operators worldwide.

Figure 14. RFID tags on letter trays and Real-time load control

Key Benefits:

- Better utilization of postal letter trays.

- Possibility to analyze though-put times of mail and letter trays at distribution centre.

- Knowing the location of trays improves their availability throughout the entire logistics chain.

- Knowing the location and contents of trays improves the possibility of managing the tray sorting process right on time.

- On automatic handling systems, such as tray sorters, the reading rate can be improved dramatically compared to that of bar codes - reducing manual intervention.

- Being able to identify trays helps to improve service and maintenance.

4.6.4. Mail bag tracking

Mail bags are widely used all over the world for transporting letters. The use of the mail bags differs between postal operators from transporting standard letters, to added value letters or to being used in closed customer loops. Each mail bag has a passive RFID tag that contains information about letters, which are inside the bag and some other additional information useful for sorting and other postal processes (Figure 15).

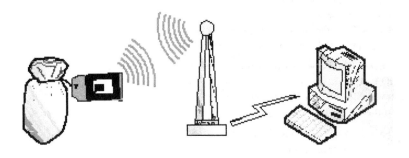

Figure 15. RFID tag placed on Mail Bag

Independent of how each postal operator is using the mail bags, tracking them can improve their competitiveness by means of:

- Optimization of processes.
- Optimization of routes.
- Control of Quality of Service.
- Internal documentation of handovers.
- Customer documentation.

Tracking solution is based on:

- Bar codes.
- Active RFID tags.
- Passive RFID tags.
- Hand-held terminals or PDAs.
- Automatic scanners.

5. The impact of the operational characteristics on the readability in postal sector

In this part we show the reality of using RFID technology to identify the letter by specific analysis of the legibility of letters in the crate. The goal was to assess whether it is possible to achieve 100% legibility of letters stored in the crate using a postal RFID technology.

To determine the success of reading measurements were performed on letter mail stored in the actual postal crate using the RFID reader and two antennas from Alien, label affixed to objects and middleware management program. Under review was to create RFID systems and perform test measurements to evaluate the success of the load of letters stored in crates and stored the measurements are properly presented and evaluated in the framework to create web application related to middleware program that is designed to manage RFID reader.

For the purposes of measurement was the technical background of Alien - RFID reader, RFID tags, and two antennas, which was created by the RFID gate. Used middleware program provided by the Italian company Aton, s.p.a. web application was developed in an environment with a PHP MySQL database system. Principle of RFID technology is as follows:

- the base of the system is reading device (reader) RFID systems and serves as a transmitter and a receiver of radio waves

- part of the reader are one or more antennas through which the reader is able to transmit electromagnetic waves to a radiofrequency, and transmit the encoded information,

- using RFID transponder tag is received electromagnetic waves with information encoded converted into an electric charge is stored on RFID tags,

- transformation of electromagnetic waves into an electrical charge is possible that the RFID tag is able to broadcast their own radio waves with its own unique encoded information,

- reader receives the signal modulated with disabilities. The information thus obtained is further processed and sent to the superior information systems.

5.1. Orientation and location of identifier

Identifiers are polarized as well as antennas. For optimal performance RFID read range and the polarization must be parallel to the polarization of the antenna. For most of the current is the polarization parallel to the longer side. Ideal antenna alignment and location identifier is an identifier in front of the antenna and the longer side oriented parallel to the polarization of the antenna. Real but it is virtually impossible to guarantee. In all applications, but it is important to align the antenna with the antenna system identifier reader. Same alignment orientation identifier in phase with the direct model antenna returns optimal results. However, the general rule is that the identifier may be disoriented by about 15 ° angle in any direction with negligible performance degradation. Correct adjustment of the system may allow an even greater tolerance. This tolerance to disorientation system allows you to read the label orientation and angle of presentation changes depending on their trajectory through reading.

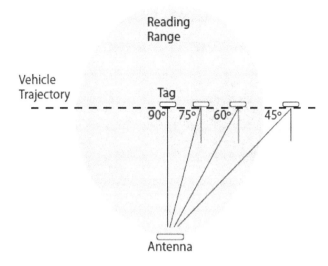

Figure 16. This is illustration shows the identifier transmitted by antennas around the reader.

This illustration shows the identifier transmitted by antennas around the reader. As shown in figure 17. range reading is weaker if the identifier to a greater angle to the antenna.

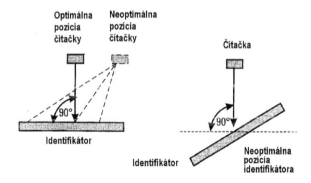

Figure 17. The optimal position of antenna and tag (identifier)

Reading range may be affected by pitch, roll or diverting of identifier. In a further assume that the antenna polarization is parallel to the long side identifier.

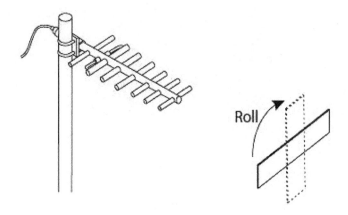

Scheme 1. Rotation (roll) the identifier in a clockwise or counterclockwise direction will cause loss of performance. This loss increases with expanding angle of rotation - the optimum approach angle is 90 °. That may be why the orientation of the identifier used to avoid reading other remote signals from any recources.

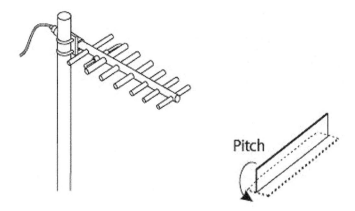

Scheme 2. Inclination or tilt (pitch) of identifier - the rear rotation (*front to back*) around a horizontal axis) affects performance only slightly.

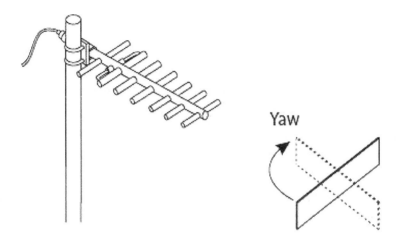

Yaw

Scheme 3. Rotation (yaw) facing each other (end on end) about its vertical axis is for further consideration. As the angle of identifier rotation increases from the antenna, area identifier's internal antenna, which is within the reader field is shrinking. With this reduction is also associated identifier readability.

Differences in surface mounting, the angle and height of placement, as well as changes in the angle of the transition identifier reading area are compensated if allowed a sufficient margin for optimal orientation and alignment

5.2. Evaluation of test success

The main purpose of making measurements and the creation of applications was to evaluate the success of the measurements and draw conclusions about the most significant and important impacts that affect the success of reading. The effects have been studied for measurements are as follows:

- save correspondence with the measurements
- the use of crates,
- free storage correspondence,
- number of measured items.

5.2.1. Impact of the imposition of letters

During performing measurements correspondence can be letters stored in several ways to create a number of test locations. Basis to build positions in the imposition items vertically or horizontally as shown next figure.

At horizontally aligned storage correspondence can be created as the following positions:

- items stored horizontally narrow side facing the antenna,
- items stored horizontally wider side facing the antenna,
- items stored horizontally, in the gate rotated 360 degrees.

At a vertical aligned correspondence can be created as the following positions:

- Items stored vertically, towards the flat antenna
- items stored vertically, perpendicular to the flat antenna
- items stored vertically in the gate rotated for about 360 degrees.

Other than those specified positions were also carried out tests on unaligned shipments. At vertical alignment shipment is possible to distinguish whether the letter is placed RFID tags glued upward or downward. Depicting the imposition of letters is shown in the following table. In order to determine the optimal storage correspondence tests were performed in all these positions.

Description deposit of letter items	Reading success (%)
Aligned vertically toward the surface of the antenna, the tags below	29,03
Aligned vertically toward the surface of the antenna, the tags above	64,98
Aligned vertically, perpendicular to the flat antenna, the tags below	4,15
Aligned vertically, perpendicular to the flat antenna, the tags above	11,98
Horizontally aligned, wider side facing the antenna	26,27
Horizontally aligned, narrow side facing the antenna	11,06
Aligned vertically, rotate the gate about 360 degrees	65,44
Aligned horizontally, rotate the gate about 360 degrees	42,40
Misaligned, randomly placed	85,25

Table 1. The impact of the imposition of letters

As shown in table above the highest percentage was reached at a loading unaligned accidentally saved letter. Saving is but random, and in greater numbers there is no guarantee that the RFID tags do not overlap more than one shipment. We can assume that for larger numbers, this percentage declined.

When comparing the measurements of success with storing correspondence RFID tags up and save measurements made with RFID tags can be seen down a significant difference. Greater success is achieved when depositing RFID tag upwards, which is due to greater freedom for the RFID tags. Large differences are visible when you turn the leaf surface shipments towards RFID tag antenna. Compared to the stored correspondence surface perpendicular to the RFID tag antenna is the difference in the success of loading more than 50%.

Measurements were carried out with the type of gate in which both antennas are on the sides. Gate type significantly influenced the success of horizontal loading and shipments compared to a vertical were significantly lower. Rotate the trays in the gate 360 degrees slightly increased readability vertically or horizontally stored correspondence. Optimal solution in terms of deposit of letters on the measurements is to store:

- vertical,

- flat plate toward the RFID antenna,

- Implementation in gate turned 360 degrees.

5.2.2. The impact of the use of containers

To determine the impact of the imposition of letters in the crate measurements were not made only in the crate, but also in bulk correspondence without containers. With the settings and save the items in the same position was achieved the following results:

	Reading success (%)	
Imposition of letter items	use crate	without crate
aligned, horizontally placed	82,95	84,79
aligned, vertically, tag surface to side antenna	91,71	63,59
aligned, vertically, tag surface upright to side antenna	53,46	46,08
aligned, vertical rotation	90,32	80,18
aligned, horizontally rotation	83,41	93,09
unaligned, random stored	98,16	96,31
overall	83,35	77,34

Table 2. The impact of the use of containers

The table shows that the use of containers has not a significant impact on the success of improvement or deterioration reading of letter items. When using containers to store the correspondence is achieved even greater average success on reading. This is probably due to the freer depositing correspondence in containers than in the same position simulations without containers, especially in the upright position. At horizontal position, where it was easy to simulate the same position was achieve slightly higher readability without the use of crates.

5.2.3. The impact of free scope of stored letter items

By examining the different variations of the deposit of letter mail has proved an important factor affecting the success of slackness between RFID tags glued to the letter. It was had done testing with the following settings of slackness between the letter post:

- separate correspondence by carton;
- the bulk correspondence (classical)
- letter correspondence pressed together.

Degree of freedom between the letter items	Successful reading (%)
pressed together	68,66
stored slackness	69,59
separate by carton	100,00

Table 3. The impact of free scope of stored letter items

The table shows that the separation of the carton shipments has a significant impact on the success of loading achieved is 100% readable. Crushed shipments only slightly worse compared to the readability of bulk shipments.

5.2.4. The impact of other elements

Unit shipments for the cardboard several measurements were carried out to confirm 100% readability:

Studied impact	Successful reading (%)	
speed of running by gate	4 sec.	100,00
	2 sec.	100,00
	quickly	100,00
antenna distance	80 cm	100,00
	60 cm	100,00
antenna intensity	90%	100,00
	75%	100,00
	60%	100,00
	40%	100,00
	20%	95,85
optimal storage of letter		97,24

Table 4. The impact of other elements

By separating mail boxes are reaching nearly all settings by 100% readable. Mild impairment occurred only at very low intensity at 20% and save correspondence area perpendicular to the antenna. But even in these cases is very high loading percentage. But the question remains questionable real use in practice.

5.3. Model 1 – Evaluation of the feasibility of using RFID technology

At real-saving correspondence to crates are stored in the manner:

- aligned vertically,

- unseparated to each other as shown next figure.

Figure 18. Postal crate

By a given type of deposit correspondence, the maximum load value of the success achieved in the following settings:

- **intensity** of the RFID reader to a maximum value,

- **type of gate** - one antenna on the side, a top antenna,

- **remain** in the gate at least 2 seconds,

- **freely** save letter items.

After several performed in those settings with 31 letter items were obtained the following results:

Measurement count	Number of reading items	Successful reading (%)
1	30	96,77
2	29	93,55
3	30	96,77
4	30	96,77
5	24	77,42
6	25	80,65
7	31	100,00
Overall	199	91,71

Table 5. Results of measurement

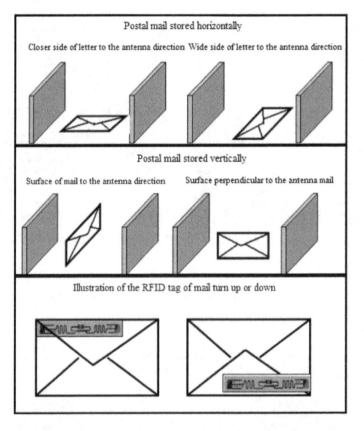

Figure 19. Position of mail while reading

At horizontally stored correspondence, following positions can be created:

- items stored horizontally narrow side facing the antenna,
- items stored horizontally wider side facing the antenna,
- items stored horizontally, in the gate rotated in 360 degrees.

At a vertically stored correspondence, following positions can be created:

- items stored vertically, surface towards the antenna
- items stored vertically, surface perpendicular to the antenna
- items stored vertically in the gate rotated in 360 degrees.

Except for those specified positions, tests were also realized on unaligned shipments. Within vertically aligned shipment, it is possible to distinguish whether an RFID tag is placed upward or downward on letter. Depicting the imposition of letters is shown in the following table.

In order to determine the optimal storage of correspondence, the tests were performed in all mentioned positions.

Description deposit of letter items	Success of reading (%)
Aligned vertically toward the surface of the antenna, tags placed on bottom	29,03
Aligned vertically toward the surface of the antenna, tags placed on top	64,98
Aligned vertically, perpendicular to the flat antenna, tags placed on bottom	4,15
Aligned vertically, perpendicular to the flat antenna, tags placed on top	11,98
Horizontally aligned, wider side facing the antenna	26,27
Horizontally aligned, narrow side facing the antenna	11,06
Aligned vertically, rotate the gate about 360 degrees	65,44
Aligned horizontally, rotate the gate about 360 degrees	42,40
Misaligned, randomly placed	85,25

Table 6. The success of reading the different position of mail

5.4. Model 2 – Test of readability of postal items through the RFID technology

One of the methods that could significantly make the process of identifying postal items in transport condition more effective is above mentioned RFID technology. As a wireless technology, without visual contact with the shipment, it tracks and identifies the contents without the need of manual handling from the crate. This allows easier and more efficient handling of supporting documents (creating the list of items, checking the presence of item) of postal sacks/bags and containers. With regard to price and the quantity of items processes, a question arrives: Is RFID technology effective and should be used for all shipment, including letters? As already mentioned - due to the large quantities of letter items and still quite high price of RFID tags - the method could be appropriate only for registered letters/mail. The actual implementation design of RFID technology, as shown in Figure 3 could be divided into the following phases:

1st phase - tracking between the HSS

2nd phase – tracking between the HSS and OSS.

3rd phase – tracking between the OSS and final post office (point of delivery).

Because of our basic interest is in the RFID technology we tried to test of readability RFID tags placed on postal items in various situations. Basic assumption is the use of RFID gates at the entrance and output to the processing unit as show next figures.

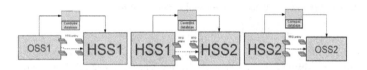

Figure 20. RFID gates at the entrance and output to the processing unit

In order to verify the practical applicability of this technology we have dealt with the preparation and implementation of practical activities through which we examined reading RFID tags. The object of these measurements was to determine the statistical characteristics of reading success and reading passive tags, placed on postal items, located in the mail bag. The aim was to provide sufficient information as accurately measured under different conditions that can occur in a real situation, including a draft measure, which would lead to the improvement of measured data.

Therefore we try to simulate a real postal process in conditions close to operational and test this technology on next component set configuration:

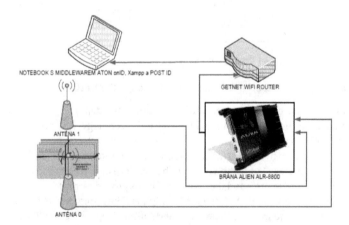

Figure 21. Principle component links

When some bundle or bundles placed in a bag entry into the detection field begin to transfer the identification data from the RFID tag to the antennas of gate. The gate extends the data (by adding date, time, number of particular antenna,…) and send them to system. Thus processed data are transmitted through the wifi router middleware Aton onId into notebook.

Figure 21 (figure located in next section) describes the principle of software components and their cooperation - communication between web applications POST ID, MySQL database server and middleware onid Aton.

5.4.1. Description of the model

There was used a software from Italian company Aton, also known as middleware, which provides the management, organizational and communication operations between different applications. In our case, the firmware Alien Gate and other applications, particularly database server. Onid Aton itself is not monolithic program, but it is a functional connection Java service console (java server) and the graphic manager called Qflow. Itself Qflow intuitive and easy enabled an interactive creation and administration of custom processes.

Major elements are program elements, called the processor to implement elementary operations (reading from the gateway, filtering, record the output, etc..)

- The first step is to enter the configuration data to POST ID. From there shall be deposited directly into database tables. The subject of this storage is data on the number of configuration items and numbers.

- In a second step, after the start of broadcasting alien element and their detection by InlineProcesor made load measurement numbers, the number of items and the configuration number and attach it to information from the antennas.

- In third step, the data are extended by the information about time and date using TimeFormatter processors. The first two into generators of text and xml files with a resolution by the uniqueness of the registration data. The third way into InsertProcessor, where the data are entered into the database. Fourth way turns itself to LackEvents processor. In the case that in a defined time there is here not recorded any new message from the gateway, it sends a new message to next two processors, which on the basis of the received message (MessageGenerator) to increase the value of measurement number by 1 and this value by updating the database InsertProcessorA.

- The second CommandExecutor processor on receipt of a report by running the alarm indicates the new number of measurements. The measurement consists of setting values in POST ID and physical adjustment of antennas. The effort was to make sure if it was possible the most accurate and smoothest possible transition from the beginning to the end of the runway. After making the transition waiting for the time needed for detection of zero, which means the CPU and LackEvents CommandExecutor will sound, indicating the end of measurement and readiness for the next measurement. At the same time processor MessageCounter increased number of first measurement after finishing the sound detection is possible again to make the switch between the antennas to the selected track.

Full application part is shown in Figure 22.

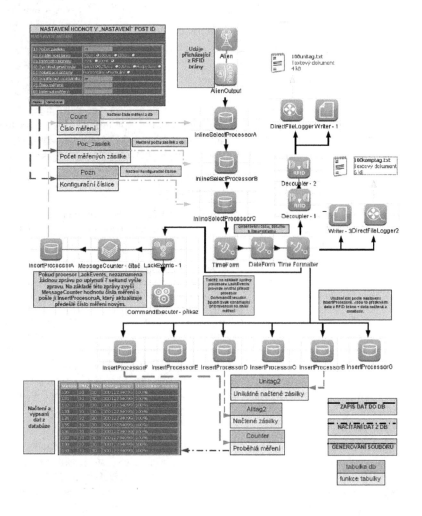

Figure 22. Final configuration model based on ATON onID

5.4.2. Test of readability of postal items

Measurements carried out in an improvised laboratory in the premises of the computer lab of the University of Žilina. There were measured passive tags placed uniformly on all mail in the middle of the upper left corner. Tags were placed so strictly because simulate challenging situation that could occur in real practice, so that all shipments under the labels overlap, the close neighbours. This arrangement could cause the EM waves emitted by RFID tags will interfere with each other. For each item was then transcribed RFID tag number and serial number marked for later processing easier statistical information. The object of measurement items were deposited into the mail bags, which are grouped into a bundle. To determine the characteristics of reading and expanding sub-measure was introduced by another character, and that is the position of the beam due to the antennas. These positions are defined (according to Figure 23):

1. boundle horizontally - the length of the area enclosing antennas,

2. boundle horizontally - the width of the area enclosing antennas,

3. boundle vertically - party address shipments parallel flat antennas,

4. boundle vertically - mail address side perpendicular to the plane antenna.

Likewise, in our measurements were sequenced according to the serial number of items, grouped into bundles, according to the size of the consignments as shown in Figure 23.

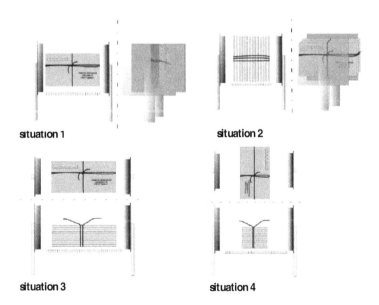

situation 1   situation 2

situation 3   situation 4

Figure 23. Configurations of letters bundles

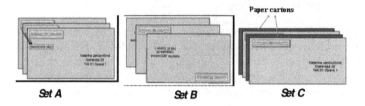

Set A　　　　　　　　**Set B**　　　　　　　　**Set C**

Figure 24. Several set of letter bundles

We have used two ways of transporting items via gate:

1. Static transfer through the postal truck or conveyor, also examined the transport unit volumes, which are in relative peace in terms of positioning items

2. Dynamic hand, respectively manual transfer using human power - move with a rate shocks, which could help to better read the labels in bundles.

All data recorded after the measurement time was subject of evaluation, and because of the large scale of the recorded data was be evaluated only average and cumulative results Determining the accuracy of measurements based on the statistical characteristics - it is a statistical description, which expresses the degree of statistical variability of the file, it indicates the letter R, It indicates the difference between the largest and smallest value and in some extent we are able to denounce both the large inaccuracies in the measurement occurred. It is expressed by the formula $R = X_{max} - X_{min}$. In percentage terms inaccuracies modified formula looks as follows:

$$Z = ((X_{max} - X_{min}) / number\_of\_items) * 100$$

Based on this formula was compiled by chart positions inaccuracies sets.

Since the evaluation of this quantity of data with the graphic processing is substantially opaque (a sample can be figure 25 with a graphical evaluation, which is a preview of kits depending on the speed of transition between the antennas) and it is not possible to present all the results of measurements on such a small space of this contribution will sum up only the basic results of the measurements and focus only on some important findings.

The measurement is made clear that that some parameters are irrelevant in terms of readability, such as speed of shipments run through the transition zone readers are relatively independent (readability in an average of about 76% to 2% deviation, the readers distance is taken with a 77% deviation around 6% use conveyor with 80% deviation around 4%, or manual switch (94% with a deviation of about 2%.

It is interesting that in an evaluation of readability situations of the consignment given the readers runs through the gate (table 1 upper), in some cases is sufficient and relatively uniform (situation 2 and 3 value of 100%), while what for example situation 4 is the readership in wide range of 2% up 92%.

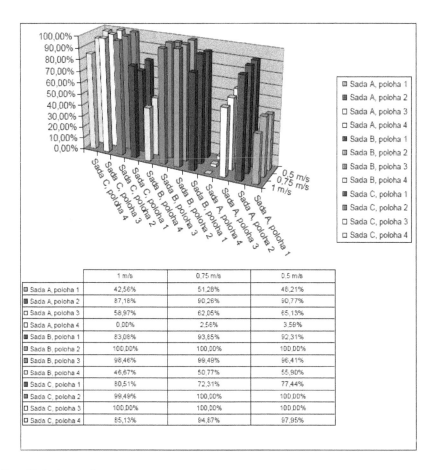

	1 m/s	0,75 m/s	0,5 m/s
▣ Sada A, poloha 1	42,56%	51,28%	46,21%
▦ Sada A, poloha 2	87,18%	90,26%	90,77%
☐ Sada A, poloha 3	58,97%	62,05%	65,13%
☐ Sada A, poloha 4	0,00%	2,56%	3,59%
▦ Sada B, poloha 1	83,08%	93,65%	92,31%
▣ Sada B, poloha 2	100,00%	100,00%	100,00%
▦ Sada B, poloha 3	98,46%	99,49%	96,41%
☐ Sada B, poloha 4	46,67%	50,77%	55,90%
▦ Sada C, poloha 1	80,51%	72,31%	77,44%
▦ Sada C, poloha 2	99,49%	100,00%	100,00%
☐ Sada C, poloha 3	100,00%	100,00%	100,00%
☐ Sada C, poloha 4	85,13%	94,87%	97,95%

Figure 25. Final test results

Set	situation 1	situation 2	situation 3	situation 4	average
A	47,35%	89,40%	62,05%	2,05%	50,2%
B	89,74%	**100,00%**	98,12%	51,11%	84,7%
C	76,75%	99,83%	**100,00%**	92,65%	92,3%

Table 7. Basic dependency between sets of readability and situations

The overall success of the method of transition as the distance of antennas for different speed ranges from 81% - 87%, and has a major impact on readability, similar to the way the transition between sets of antennas is relative stable (81% to 95%)

Based on the evaluation of measurement data cannot be identified unambiguously exclude or recommend the use of this technology in practice. These measurements may be partly conditional on imprecision caused by a provisional Laboratories. There is unable to clearly provide the desired stable speed and position of shipments due to the antenna. The end result is therefore a lack of readability of RFID tags in a traditional way-now commonly used in practice in the post measurement known as the set A. Although in other cases, the readability is very high and almost 100% (set C or B), there were other aspects that significantly affect its use.

6. Conclusion

RFID technology is still growing up and there is several type of application, which you can use in condition of the postal processes. This chapter deals with type of application that are common uses in postal sector such as mail bag, letter trays, roll cages and vehicle tracking and managing application. All these application are useful for the track and trace system and it presents added value for costumers. Most postal services provide at least a limited form of track and trace, particularly for premium delivery services. Today, tracking uses bar codes. Switching to RFID tags can lower tracking labor by eliminating the need for most manual piece handling. RFID is a very useful and exciting technology. It seems that everywhere one looks there is some article about RFID and the huge benefits its technology promises. Moreover, there are many examples that demonstrate how this technology is fulfilling its potential.

Based on the measurements it can be concluded, with some exceptions that prove the rule, the closer they are to each antenna, the greater the success of reading RFID tags. Given the large dispersion of values it can be concluded that some elements are simply eliminated They can, for example using multiple counting gates, respectively antennas (eliminating the position of shipments), or the use of such specialty (bubble) envelopes for magnification air gap between consignments (as by set C)

This article deals with identification of postal items and transport units in logistic chain of postal operators. It described scheme of the transport process, including planned technology and there is also simulated a real postal process in conditions close to operational. Article is part of the projects described below, which, together with the afore-mentioned application, will improve the learning process at the Department of Communications.

The benefits of RFID technology can be reaped if RFID events give realtime visibility to the business processes either already in place or to new ones. The backend systems give a business context to the RFID events collected from the RFID data collection tools and then invoke the right business process in real time (or near real time). Protecting the backend system is vital from the various security threats at the network level (attacking ONS or network communication between data collection tool and backend system) or at the data level (spurious events).The network level attacks can be prevented by using secured communications between various processes. The data attacks are hard to deal with, and application designers must take special care to differentiate spurious events from good events and then act on the good ones almost in real time. Since data is collected using automated data collection techniques, application designers must clean the repository where good RFID events are stored.

Costs of the security regarding RFID technology got implemented in a company's infrastructure still presents relatively expensive attribute in the eye of CEO's. Although the price of active and passive tags is still being reduced and RF technology becomes continuously more and more popular in a field of logistics, supply chains, toll systems, postal services, retailers or asset management, the relevance has to be put on a confrontation of costs of RFID implementation and its explicit use towards the eventual probability of attack.

In the relation on main aim of this article was focus on options of implementation mobile technology includes RFID technology into postal transport area. The content of introductory chapter is approximation the best terms of category and theoretical knowledge, which is used in this article. Further there are characterized the postal transport network, individual transport units, which is use in this area.

By optimal settings in real conditions the average maximum reading percentage is about 91.71%. Significant effect to increase the readability, RFID tags have been pointing toward a flat antenna and the antenna by sensing the top. Very significant impact on increasing or reducing the readability of the number of letters had the crate. For the purposes of measurement was used in 31 letters. The real use of the crate contained a much higher number of letters, which would likely significantly reduce readability.

Based on these results, identification of letters in the crate, RFID technology is not yet, given the technical conditions for real. Achievements, however, were relatively high. Some uncertainties should be possible to eliminate appropriate technical configuration (number of gates and Antennas, their location, etc.). On the other side some, particularly the operating elements (separation of cardboard) can be solved as special packaging elements (bubble envelope to increase the air gap), etc.) can ensure a desired level of reliability required reading. Therefore, we expect that further testing with a larger number of antennas other types of antennas, readers and RFID tags in our AIDC laboratory.

Acknowledgement

This work was supported in part by grant and research project VEGA 1/0421/12 - Modeling diffusion of knowledge in business value chains (60%), Centre of excellence for systems and services of intelligent transport II., ITMS 26220120050 supported by the Research & Development Operational Programme funded by the ERDF (20%) and institutional projects: 3/KS/2012 - Supporting education through educational multimedia applications [10%] and 1/KS/2012 - Sensitivity analysis of contact points to the costs arising from the provision of UPS [10%]

Author details

Juraj Vaculík, Peter Kolarovszki and Jiří Tengler

*Address all correspondence to: juvac@fpedas.uniza.sk

University of Zilina, Faculty of Operation and Economics of Transport and Communications, Department of Communications, Slovakia

References

[1] Hunt V. D., Puglia a. , Puglia m., RFID: A Guide to Radio Frequency Identification, ISBN-13: 978-04701072007.

[2] Jones, ERIC, Chung, CHRISTOPHER : RFID in Logistics: A Practical Introduction, ISBN 978-0849385261, 2007

[3] Finkenzeller, K.: *Fundamentals and Applications in Contactless Smart Cards and Identification.* In: RFID Handbook. John Wiley & Sons, 2006. ISBN 0-470-84402-7.

[4] LEHPAMER, H.: *RFID Design Principles.* Artech House, Inc, Norwood, MA, USA, 2008. ISBN 978-1-59693-194-7.

[5] Myerson Judith M., *RFID in the Supply Chain: A Guide to Selection and Implementation,* Syngres, 2007. 215 s. ISBN 978-08493301086.

[6] Poirier Charles, RFID *Strategic Implementation and ROI : A Practical Roadmap to Success,* J. Ross Publishing, Inc., 2006. 243 s. ISBN 1-932159-47-9.

[7] Švadlenka L. a kol. Transportation and communication system, První vydání. Pardubice : Univerzita Pardubice, 2006. 136 s. ISBN 80 – 7194 – 911 – 6

[8] ŠVADLENKA, Libor. RFID in postal and courier services. In PRASAD, B. V. S. , KALAI , Selvan. Supply chain management in services industry : an introduction. 1st edition. Hyderabad (India) : Icfai Books, 2007. s. 68-74. ISBN 81-314-0756-X.

[9] Thornton F.,Haines B., Bhargava H., *RFID Security,* ISBN 1-59749-047-4, 2005.

[10] THORNTON, F. a kol.: RFID Security, Rockland: Syngres Publishing 2006. ISBN 1-59749-047-4.

[11] Zelik P.: Mobile technology in postal transport, University of Žilina, Faculty of economical and operation of f transport and communication, dissertation work

[12] Post Norway-Case Story. Located on: http://www.lyngsoesystems.com/Downloads/074.342.885-A4-Post-Norway-Case-Story.pdf, 2008

[13] Post Denmark – Case study. Located on: http://www.lyngsoesystems.com/Downloads/074.719.731-Post-DK-Case-Story-.pdf, 2008

[14] Quality monitoring solution, Located on : http://www.lyngsoesystems.com/postal/ quality_monitoring.asp, 2008

[15] *Canada Post case story.* 2008. [online]. Lyngsoe Systems (Canada). [citované 2009-03-15]. Dostupné na: http://www.lyngsoesystems.com/ Downloads/074.772.592-Canada-Post-casestory-A4.pdf

[16] Garfinkel, S - Rosenberg, B, RFID Applications, Security, and Privacy, Addison- Wesley Professional, 2005.

[17] DENNIS E. BROWN. RFID Implementation. New York (USA), McGraw-Hill Publishing, 2007

[18] Christina Soh, Sia Siew Kien, and Joanne Tay-Yap, Enterprise resource planning: cultural fits and misfits: Is ERP a universal solution? Communications of ACM, 43, 47–51, 2000.

[19] Benita M. Beamon, Supply chain design and analysis: Models and methods, International Journal of Production Economics, 55, 281–294, 1998.

[20] D.M. Lambert and M.C. Cooper, Issues in supply chain management, Industrial Marketing Management, 29, 65–83, 2000.

[21] Valerie T., Neckel W., Information technology and product lifecycle management, In Proceeding of the 1999 IEEE International Symposium on Electronics and the Environment.

Chapter 8

RFID Under Water: Technical Issues and Applications

Giuliano Benelli and Alessandro Pozzebon

Additional information is available at the end of the chapter

1. Introduction

While RFID technology is nowadays very common in many commercial and industrial sectors, from items tracking to personal identification, few studies have dealt with the chance to use RFID systems in marine or fluvial environments for underwater monitoring operations. While the technical limitations for these scenarios can be in some cases insurmountable, ad-hoc studies have proven that in some cases RFID technology can work even under water.

RFID, like all radio technologies, in unsuitable to work in presence of water. Still water is not a natural conductor, but the presence of dissolved salts or other materials turns it into a partial conductor. Electromagnetic waves cannot travel through electrical conductors: this means that in most cases radio waves cannot be used to communicate under water. Anyway, studies have proven that the chance to transmit radio signals under water mainly depends on two factors: the conductivity of water and the frequency of the radio wave. While the conductivity of water is a factor that cannot be modified to increase the possibility to use radio waves under water, the only factor that can be modified to increase the performances is obviously the radio frequency.

This factor has already been employed when using the electromagnetic fields for the common radio transmissions: Very Low Frequency radio waves (VLF – 3-30kHz) have proven to be able to penetrate sea water to a depth up to 20 meters, while Extremely Low Frequency radio waves (ELF - 3-300Hz) can travel in sea water up to hundreds of meters. Anyway, these frequency bands present severe technical limitations. First of all, their extremely long wavelengths require antennas of very big dimensions: frequencies lower than 100Hz have wavelengths of thousands of kms, forcing to use antennas covering wide areas. Secondly, due to their narrow bandwidth, these frequencies can be used to transmit only text signals al slow data rates.

Some of these considerations can be applied also to RFID systems. First of all the use of active technologies is discouraged by many factors: at lower frequencies only passive systems can be found; moreover, the use of active systems is also impeded by the required dimensions of the antennas. Due to these limitations, only two RFID technologies can be employed for underwater applications: the High Frequency systems, operating at 13.56MHz and the Low Frequency systems, operating in the 125-134kHz band. The first solution (13.56MHz) still presents some severe limitations due to the reduction of the reading range: with common desktop antennas the reduction in the range is up to 80%, forcing to bring the transponder practically in contact with reader antenna. For the second solution (125-134kHz) the reduction is lower (around 30%) and the reading at a distance is still achievable. Laboratory tests proved that, with long-range antennas, a 50cm reading range is still achievable.

Both these two solutions can be anyway employed to set up RFID systems working in under water environments. Some solutions can already be found in some parts of the world [1]. USS Navy is testing the use of RFID technology for their applications based on the use of Unmanned Underwater Vehicles. Other applications foresee the use of RFID for the monitoring of underwater pipelines, with RFID transponders employed as markers to guarantee the integrity of the pipes. RFID has also been employed in aquariums to identify fishes, in the same way as Low Frequency RFID capsules are employed in cattle breeding. Finally RFID has been employed as a way to track the movement of pebbles on beaches, in order to analyse the impact of coastal erosion during sea storms.

The chapter will be subdivided in four main sections.

In the first section, the transmission of radio signals in water will be analysed. Details will be given on how the presence of water affects the electromagnetic fields, and examples of applications working in the VLF and ELF bands will be provided.

The second section will focus only on RFID. Technical data will be provided concerning the signal attenuation due to the presence of water. Some results will be given to prove the agreement of experimental data with the theoretical analysis.

In the third section the state of the art concerning under water RFID applications already existing all around the world will be provided. The few already tested applications will be described in detail.

Finally, in the fourth section some future applications based on this technology will be proposed.

2. Underwater radio signals

2.1. Water electric and magnetic properties

Water molecule is composed by two oxygen atoms and one hydrogen atom bonded together by a covalent bond. Oxygen has a negative charge, while the two hydrogen atoms have a positive charge: this means that the vertex of the molecule has a partial negative charge, while the two ends have a partial positive charge. A molecule with such a charge equilibrium is called electric dipole, and is characterized by its dipolar momentum μ, defined as the product between the absolute value of one of the two charges and the distance between them. This value indicates the tendency of a dipole to orientate under the effect of a uniform electric field.

While still water has a very low electrical conductivity, this value increases in presence of ionized molecules, in proportion to their concentration. When a salt is melt in still water, the single molecules are equally perfused in the whole liquid so that each single volume portion of the solution dissociates, creating many positive and negative ions that remain in the solution together with all the other molecules that aren't dissociated. This phenomenon is called electrolytic dissociation, and the so created solutions are called electrolytic solutions. These solutions can be crossed by an electrical current, in contraposition with still water that acts as a pure insulator.

2.2. Marine water

The chemical composition of marine water is influenced by several biological, chemical and physical factors: one simple example is the presence of rivers that add every day new chemical materials to the water. On the other side, other materials are removed by the action of organisms and due to erosion. Anyway, the most part of the salts dissolved in marine water remains almost constant due this continuous interchange phenomenon. The most important factors that influence the chemical composition of the marine water are the following:

- The draining of materials deriving from human activities;

- The interaction between the sea surface and the atmosphere;

- The processes between the ions in solution;

- The biochemical processes.

The elements that can be found is marine water are around 70, but only 6 of them represent the 99% of the total. These predominant salts are:

- Chloride (Cl): 55.04 wt%

- Sodium (Na): 30.61 wt%

- Sulphate (SO_4^{2-}): 7.68 wt%

- Magnesium (Mg): 3.69 wt%

- Calcium (Ca): 1.16 wt%

- Potassium (K): 1.10 wt%

The symbol (wt%) stands for the mass fraction, and represents the concentration of a solution or the entity of the presence of an element in a solution. The quantity of these ions is proportional to the salinity of water, a parameter describing the concentration of dissolved salts in water. Due to the evaporation, this value is lower at the poles (around 3.1%) and higher at the tropics (around 3.8%), with the highest value for an open sea reached by the Red Sea (4%, with a peak of 4.1% in the Northern parts). Moreover, salinity is lower close to the coasts due to the inflow of fresh water by the rivers. Salinity affects the conductivity of water: while this parameter also depends from the water temperature and pressure, it ranges from around 2 S/m to around 6 S/m. Anyway, in most cases it can be considered constant, with a value of 4 S/m. Water is then a conductor.

Once the value of water conductivity is known, it can be used to calculate the values of the penetration depth and of the attenuation.

The penetration depth δ is the distance where the electrical and magnetic fields are reduced of a $1/e$ factor, and it can be calculated using the following formula:

$$\delta = \frac{1}{\sqrt{\pi f \mu \sigma}}\ m$$

where f is the frequency of the electromagnetic wave, μ is the absolute magnetic permeability of the conductor and σ is the conductivity. While water is a diamagnetic material, their absolute magnetic permeability can be considered the same as the vacuum magnetic permeability, i.e. $\mu_0 = 4\pi*10$ H/m. This means that, with the conductivity considered constant, the penetration depth only depends on the frequency: the higher is the frequency, the lower is the penetration depth.

The attenuation α can be calculated using the following formula [2]:

$$\alpha = 0.0173\sqrt{f\sigma}\ dB/m$$

where f is the frequency of the electromagnetic wave and σ is the water conductivity that, as said before, can be considered constant. Attenuation is then in inverse proportion with the frequency and then obviously also with the penetration depth.

2.3. Fresh water

Around 97% of the water of the world is found in seas and oceans, while two thirds of the remaining 3% of fresh water is retained as ice in glaciers and at the poles. This means that the most part of studies that can be found concerning the chance to communicate under water using the electromagnetic fields focuses on the marine environment.

Anyway, similar considerations as the ones made for salt water apply to fresh water. The biggest difference derives from the different values of salinity that are detected in fresh water. While the salinity of salt water is around 3.5% (See section 2.2), in fresh water this value decreases down to 0.05%. Anyway, unlike marine water, a general analysis concerning the quantity and typology of salts that can be found in fresh water is impossible to carry out due to the single peculiarities of rivers, lakes, and the chemical and geological composition of the territories that they pass through and where they are located.

A different value in salinity also means a different value in conductivity. In particular, conductivity of fresh water ranges from 30 to $2000 \mu S$/cm: these are nevertheless extreme values; river water conductivity usually ranges from 50 to $1500 \mu S/cm$, while rivers supporting a good wildlife usually range from 150 to $500\mu S/cm$. This value is notably lower than the average one for marine water. The main consequence of this fact is that for fresh water the penetration depth is higher and the attenuation is lower.

2.4. Underwater radio communication

Some easy calculations prove that the electromagnetic fields can be used to transmit radio signals under water (Especially under the sea) only when their frequency is very low. As an example we can calculate the penetration depth for an electromagnetic wave traveling through salt water at frequency of 10kHz, using the average values for μ and σ:

$$\delta_{10kHz} = \frac{1}{\sqrt{\pi f \mu \sigma}} = \frac{1}{\sqrt{\pi \bullet 10^4 \bullet 4\pi \bullet 10^{-7} \bullet 4}} \approx 2.5m$$

This value allows a short range communication, while long range communication requires even lower frequencies.

Looking at fresh water the situations is a little bit better. The previous calculation can be made, using a very low conductivity value of 30 $\mu S/cm$ (3mS/m):

$$\delta_{10kHz} = \frac{1}{\sqrt{\pi f \mu \sigma}} = \frac{1}{\sqrt{\pi \bullet 10^4 \bullet 4\pi \bullet 10^{-7} \bullet 3 \bullet 10^{-3}}} = 92m$$

Anyway, while this value is higher, long range communication is not allowed when the operative frequency is higher than some kHz.

As a consequence of the previous analysis, the only bands that have been used for underwater radio communication have been the ELF (Extremely Low Frequency) band, ranging form 3 to 300 Hz, with the sub-band ranging from 30 to 300 Hz called SLF (Super Low Frequency) band, and the VLF (Very Low Frequency band).

The ELF band was used for the communication with submarines both by the US and the Russian Navies. The US system, called Seafarer, operated at the frequency of 78Hz, while the Russian one, called ZEVS, operated at the frequency of 82Hz. These systems had a penetration depth in the order of 10km, allowing thus a communication from a fixed station on the sea surface with a submarine traveling close to the ocean floor. Anyway, the realization of a communication channel at these frequencies presents several technical limitations that are extremely difficult to be overcome. One of the biggest problems to be solved is the size of the antenna: its dimension has in fact to be a substantial fraction of the wavelength, but at these frequencies the dimension of the wavelength is in the order of the thousands of kilometres. The solutions found by the US and Russian Navies were complex and expensive, making prohibitive their use for civil applications.

The VLF band ranges from 3kHz to 30kHz: this means that the penetration depth of electromagnetic waves at these frequencies is in the order of ten meters. This value allows a communication with submarines positioned few meters below the sea surface. The limitations on the antenna dimensions, deriving from the big wavelength, have to be taken in account also in this case. Moreover, due to the limited bandwidth, this communication channel cannot be used to transmit audio signals, but only text messages.

3. Underwater RFID

RFID, being a radio technology, suffers from the same limitations of the standard communication channels. This means that the higher is the frequency, the lower are the chances to have a reliable communication {3-7}.

RFID systems are usually subdivided in the following bands:

- Low Frequency (LF) – 120-150kHz;

- High Frequency (HF) – 13.56MHz;

- Ultra High Frequency (UHF) – 433MHz, 868-928MHz;

- Microwave – 2.45-5.8GHz.

3.1. Salt water

As underlined in section 2, significant differences occur according as the RFID system has to be used in salt or fresh water. Starting from salt water, some calculations show that only LF RFID can be used for systems requiring a long reading distance (over 50cm). In particular at a frequency of 125kHz, the average value (Using the salinity value of 4S/m) for the penetration depth is:

$$\delta_{125kHz} = \frac{1}{\sqrt{\pi f \mu \sigma}} = \frac{1}{\sqrt{\pi \bullet 1.25 \bullet 10^5 \bullet 4\pi \bullet 10^{-7} \bullet 4}} \approx 71cm$$

This value is just lower than the maximum achievable reading range for a Low Frequency system, which is usually lower than 1m. This means that Low Frequency RFID can be theoretically used for the underwater identification of items.

Moving at higher frequencies, the use of these systems for long range identification becomes virtually impossible. The calculation for the penetration depth provides an extremely low value. Starting from the High Frequency band, where all RFID systems work at the standard frequency of 13.56MHz, with the same conditions as in the previous case, the obtained value for the penetration depth is:

$$\delta_{13.56kHz} = \frac{1}{\sqrt{\pi f \mu \sigma}} = \frac{1}{\sqrt{\pi \bullet 13.56 \bullet 10^6 \bullet 4\pi \bullet 10^{-7} \bullet 4}} \approx 68mm$$

This result proves that High Frequency RFID can be used under water only for short range solutions. In particular, due to the fact that the effectiveness of every RFID system is notably influenced by the performances of the hardware devices employed, it's possible to affirm that the chance to use High Frequency systems is limited to the applications where the tag is in close contact with the reader.

The UHF band is currently employed in many different systems and probably represents the best solution for many applications due to its good performances in terms of reading range, costs and bitrate. Anyway, its frequency is too high to allow its use also for underwater contactless applications. The calculation of the penetration depth, using an average frequency value of 800MHz (varying this value from 433MHz to 930MHz the order of magnitude remains quite constant), provides the following result:

$$\delta_{800MHz} = \frac{1}{\sqrt{\pi f \mu \sigma}} = \frac{1}{\sqrt{\pi \bullet 800 \bullet 10^6 \bullet 4\pi \bullet 10^{-7} \bullet 4}} \approx 9mm$$

This value is obviously too short to use this technical solution for other than contact applications. Only bringing a transponder in contact with the antenna of the reader, the reading becomes possible. While this fact strongly limits the possible uses of these systems, in some cases UHF systems can still become a good choice.

Finally, the Microwave band is obviously the one that provides the worst results. The value of the penetration depth is provided only for completeness, even if currently no application can be found worldwide using this technical solution:

$$\delta_{2.45GHz} = \frac{1}{\sqrt{\pi f \mu \sigma}} = \frac{1}{\sqrt{\pi \bullet 2.45 \bullet 10^9 \bullet 4\pi \bullet 10^{-7} \bullet 4}} \approx 5mm$$

Before moving to the next section a clarification has to be made. In the previous analysis no differentiation has been done on the powering method of the transponders. In fact, while active transponders usually provide higher reading ranges, they are generally used only at higher frequencies (UHF and Microwave bands): anyway, at these frequencies the penetration depth is so short that even with the most powerful active transponder no improvement in the performances of the systems would be noticeable. Moreover, even at lower frequencies, the value of the penetration depth is anyhow lower than the reading range achievable using passive transponders: therefore, a study for the use of active transponders also at these frequencies would be useless and wouldn't provide any improvement.

3.2. Fresh water

The analysis for fresh water is similar to the one carried out for salt water. The main difference derives from the fact that, while the range of the conductivity values of salt water is very short, it becomes wider in the case of fresh water. As anticipated is section 2.3, fresh water conductivity roughly varies from 30 $\mu S/cm$ to 2000 $\mu S/cm$. While both these values are notably lower than the conductivity of salt water, the differences between the obtained values for penetration depth are less distant. In order to provide an accurate set of data, the penetration depth value will be calculated both for the best (30 $\mu S/cm$) and the worst (2000 $\mu S/cm$) case.

As in the case of salt water, the analysis will begin from the Low Frequency band. In this case, at the frequency of 125kHz, with a conductivity value of 30 $\mu S/cm$ (3 mS/m), the value of penetration depth is:

$$\delta_{125kHz} = \frac{1}{\sqrt{\pi f \mu \sigma}} = \frac{1}{\sqrt{\pi \bullet 1.25 \bullet 10^5 \bullet 4\pi \bullet 10^{-7} \bullet 3 \bullet 10^{-3}}} = 26m$$

With a conductivity value of 2000 $\mu S/cm$ (0.2 S/m) the penetration depth becomes:

$$\delta_{125kHz} = \frac{1}{\sqrt{\pi f \mu \sigma}} = \frac{1}{\sqrt{\pi \bullet 1.25 \bullet 10^5 \bullet 4\pi \bullet 10^{-7} \bullet 0.2}} = 3.2m$$

Both these values are high enough to allow a reliable long range RFID communication channel.

Moving on to higher frequencies, the second evaluation is made for the High Frequency band. The calculation is made using the standard frequency of 13.56MHz. The penetration depth value with a conductivity of 30 $\mu S/cm$ (3 mS/m) is:

$$\delta_{13.56MHz} = \frac{1}{\sqrt{\pi f \mu \sigma}} = \frac{1}{\sqrt{\pi \bullet 13.56 \bullet 10^6 \bullet 4\pi \bullet 10^{-7} \bullet 3 \bullet 10^{-3}}} = 2.5m$$

With a conductivity value of 2000 $\mu S/cm$ (0.2 S/m) the penetration depth drops to:

$$\delta_{13.56MHz} = \frac{1}{\sqrt{\pi f \mu \sigma}} = \frac{1}{\sqrt{\pi \bullet 13.56 \bullet 10^6 \bullet 4\pi \bullet 10^{-7} \bullet 0.2}} \, 30cm$$

While at lower conductivity values the realization of an efficient long range RFID system could still be possible, when the water conductivity grows the penetration depth drops down to values that make this solution difficult to be implemented or even totally impossible. Anyway, the chance to use HF RFID in particular environments like rivers or lakes has to be carefully evaluated case-by-case. An additional remark has to be made: in terms of performances, LF and HF systems are similar. This means that, if the system doesn't present specific requirements, the use of LF technology is however strongly suggested.

At higher frequencies the value of penetration depth drops down to values that allow the use of these systems only for contact or short range applications. At 800MHz the penetration depth with a conductivity value respectively of 30 $\mu S/cm$ (3 mS/m) and 2000 $\mu S/cm$ (0.2 S/m) is:

$$\delta_{800MHz} = \frac{1}{\sqrt{\pi f \mu \sigma}} = \frac{1}{\sqrt{\pi \bullet 800 \bullet 10^6 \bullet 4\pi \bullet 10^{-7} \bullet 3 \bullet 10^{-3}}} \approx 32.5cm$$

and

$$\delta_{800MHz} = \frac{1}{\sqrt{\pi f \mu \sigma}} = \frac{1}{\sqrt{\pi \bullet 800 \bullet 10^6 \bullet 4\pi \bullet 10^{-7} \bullet 0.2}} \approx 4cm$$

For Microwaves, these values drop down to:

$$\delta_{2.45GHz} = \frac{1}{\sqrt{\pi f \mu \sigma}} = \frac{1}{\sqrt{\pi \bullet 2.45 \bullet 10^9 \bullet 4\pi \bullet 10^{-7} \bullet 3 \bullet 10^{-3}}} \approx 18.6cm$$

$$\delta_{2.45GHz} = \frac{1}{\sqrt{\pi f \mu \sigma}} = \frac{1}{\sqrt{\pi \bullet 2.45 \bullet 10^9 \bullet 4\pi \bullet 10^{-7} \bullet 0.2}} \approx 2.3cm$$

A remark is necessary: the values obtained for the penetration depth are ideal values and represent mainly an upper bound. This means that in most cases the effective system will present real reading ranges notably lower and in some cases it won't work at all.

	Low Frequency 125kHz	High Frequency 13.56MHz	Ultra High Frequency 800MHz	Microwaves 2.45GHz
Salt Water $4S/m$	71cm	68mm	9mm	5mm
Fresh Water $30\mu S/cm$	26m	2.5m	32.5cm	18.6cm
Fresh Water $2000\mu S/cm$	3.2m	30cm	4cm	2.3cm

Table 1. The penetration depths for the considered frequencies for both salt and fresh water

In conclusion, while theoretical data suggest that several solutions are possible when RFID is required for under water applications, it's possible to affirm that to obtain reliable results the operative frequency as to be the lowest possible. In particular:

• For salt water long range reading is obtainable only using Low Frequency systems;

• In salt water, short range or contact reading could be possible also at higher frequencies. Anyway, also in these cases a reliable reading level could be very difficult to be achieved at frequencies higher than 13.56MHz;

• For fresh water long range reading could be obtained not only with Low Frequency systems, but also with the use of High Frequency devices operating at 13.56MHz. Anyway, also in this case the use of Low Frequency is strongly recommended due to their higher reliability;

• When short range or contact reading is required in fresh water, quite all the frequencies could be efficient, even if there is a lack of studies proving the effectiveness of UHF frequencies.

4. RFID applications under water

RFID is currently one of the most widespread technologies for the automatic identification of items. There are countless fields where RFID is used for access control, items tracking, people and animal identification and many other different applications. Anyway, few applications exist where RFID is used under water.

The question of the transponders waterproofing is crucial for many applications and several devices providing a high protection level against the contact with water have been realized. Plastic tags are inherently waterproof devices, while items like wristbands have been customized to be worn also under water. Anyway, all these devices have been designed only to resist against water intrusion, and not to be read directly under water. Moreover, no reader has been realized to be used under water. Readers providing a high protection level against water can be easily found: anyway, they are designed only to be positioned on the outside, for example on building walls for access control, and then to resist against bad weather.

A step ahead is the development of transponders realized ad-hoc to be positioned on bottles or other items containing liquids. In this case the solution mainly deals with the introduction of a dielectric layer that simply separates the transponder and the liquid allowing thus its reading.

Anyway, the number of applications where the data exchange happens totally underwater is nowadays very little: the most part of these applications deals with animal tracking and environmental monitoring, mainly in marine environment.

4.1. Animal tracking

The chance to track animals, crucial for industrial stock-breeding activities, using RFID technology has probably raised for the first time the question whether is possible or not to read RFID tags immersed in water. The body of most part of living beings is mainly composed by water: as an example, around 65% of human body is composed by water. The necessity to guarantee the integrity of the tracking device (In this case the transponder) has encouraged its positioning in a place where it cannot be removed, i.e. inside the body of the animal to be tracked. While the body of the animal is mainly composed by water, to read the transponder from the outside it's necessary to find a technological solution avoiding the insulating effect of the water layer.

The use of RFID for animal tracking is nowadays very common, and has also led to the realization of two ad-hoc standards, the ISO 11784 and ISO 11785 standards, that regulate the use of RFID devices, in particular implantable transponders, for the identification of animals. Standard RFID systems for animal tracking operate at the frequency of 134.2kHz (Low Frequency band). The transponders used for this purpose are generally glass cylinder tags that are modified to be applied under the skin of the animal, to be clasped to the ear of the animal or to be ingested by the animal.

Even if these applications deal with the interaction with water, they are not properly under water systems. Anyway, RFID technology has been employed also to track animals under water. In particular, Low Frequency RFID technology has been used to identify fishes in the aquariums [8]. At the Underwater World Singapore Oceanarium, at Underwater World Pattaya, Thailand and at Virginia Aquarium & Marine Science Center, Low Frequency cylinder glass tags have been applied under the skin of a number of fishes.

The tagged fishes are identified when they come close to a long range antenna positioned on the glass of the tank where the fishes are kept. When the fish passes in front of the antenna, the identification code stored inside the transponder is read and the fish is identified. Once the fish has been identified the visitors of the aquarium can receive an interactive set of information concerning the animal. In particular, an ad-hoc software provides on a screen a picture of the fish and a description: these data are kept on the screen until a new fish passes close to the antenna.

Figure 1. The Virginia Aquarium and Marine Science antenna identifying fishes.

4.2. Pipeline monitoring

Another interesting application that foresees the use of RFID technology under water focuses on the monitoring of pipelines used to carry oil [9]. This solution has been currently only tested, while no information has been retrieved on possible effective applications nowadays working. In this kind of applications Low Frequency RFID tags were applied directly on the pipeline, keeping a fixed distance between one tag and the other.

The tags operated the frequency of 125kHz and they were customized to fit exactly on the pipe: in particular, standard Phillips Semiconductor Hitag transponders were introduced inside a protecting case, shaped on the curvature of the pipe.

Enertag, which tested the system, also developed an ad-hoc underwater reader: this was a handheld waterproof device connected with a cable to a PC positioned on a boat on the sea surface.

This system was employed to monitor the conditions of the pipeline. In practice, the transponders acted as milestones, used to identify the exact portion of pipeline. This was combined with the data concerning repairs that the pipeline had undergone, and suggesting which portion of the pipeline required assistance.

Figure 2. The Enertag system for the pipeline monitoring

4.3. Underwater navigation

US Navy analysed a possible use of RFID technology as a support for the navigation of autonomous underwater vehicles [10]. In this application tags are positioned directly on the sea bottom, and they contain information related to their position inside the area where the vehicle is moving.

The reader is embedded directly inside the vehicle: every time that a transponder comes inside the interrogating range of the reader, the information stored inside it is read and then used by the vehicle to manage its movements.

While no data has been found about an effective application of this solution, the possible uses of such a kind of system are many. Even if this solution has been proposed by the US Navy, it could be employed also in many civil applications, from the environmental monitoring to the harbour management.

4.4. Environmental monitoring

RFID technology has been used for the monitoring of coastal dynamics. The University of Siena and the University of Pisa, in Italy, have realized the so-called "Smart Pebble" system, where Low Frequency transponders are used to trace the movements of a set of pebbles along a pre-defined span of time, in order to study the dynamics of the shoreline [11].

In this system different typologies of 125kHz transponders have been employed in the last 4 years, from plastic disc tags to cylinder glass tags. These tags were inserted inside real pebbles picked up directly on the beaches where the system had to be employed: in order to allow the housing of the transponder, the pebbles were drilled. The transponder was then glued on the bottom of the small hole realized in the pebble and then it was covered with the small rocky cap extracted during the drilling operation.

Once a large set of pebbles was realized, it was positioned on the beach to be studied, following a grid pattern covering both the emerged and the submerged portion of the beach. Through an ad-hoc waterproof reader realized modifying a common reader used for access control, the pebbles were then localized after a pre-defined span of time. The starting and final positions were recorded using a GPS total station: with these data the path followed by the pebble swarm was traced, allowing geologists to easily understand the dynamics of the shoreline and the erosive effects of the meteorological events.

This application proved to be very interesting because its biggest requirement was to achieve the largest reading range possible. This constraint forced to test different hardware solutions in order to obtain the best performances especially for salt water, which was the environment where the system had to be employed. A few tests were made with HF (13.56MHz) devices but the results achieved discouraged from using this solution. In particular, the reading range obtained with a common desktop reader under salt water was lower than 3cm. This result is in accordance with the theoretical data and excludes the use of this technology for long range under sea applications.

The following experimentations were carried out on LF 125kHz systems: the theoretical analysis on this technology foresaw the chance to use them for long range applications also under sea. The tests were carried out using a long range reader usually employed for access control. Several kinds of transponders were used for the tests, from plastic discs to glass tags. The tests tried to simulate as much as possible the real environmental conditions: to achieve this result a model of the sea bottom was realized using a plastic tube. The results of the laboratory tests are shown in Table 2 and demonstrate that, using Low Frequency, long range reading is possible also under sea. Note that the experimentation was carried out in two times, and the results are then divided in two sub-sets: the first three results provide an average value from the best and worst coupling value, while the second three provide these two values separately [12]. The results are in accordance with the theoretical analysis: the achieved reading range is lower than the penetration depth, that acts then as an upper bound.

Tag Typology	Ideal Reading Range	Real Conditions
Nylon disc	55cm	41cm
ABS Plastic disc	63cm	51cm
PVC disc	49cm	36cm
Transparent disc	50cm	28-47cm
Long Glass tag (34mm)	65cm	48-63cm
Short Glass tag (14mm)	42cm	30-41cm

Table 2. Reading ranges of different Low Frequency transponders under water

The first experimentations on the Smart Pebble system were carried out in 2009 and this solution has been since then employed in several on-site applications on different beaches in Italy. The effective use of the system has roughly confirmed the results recorded in the laboratory tests: during the localization process, the transponders embedded inside the pebbles were localized even from distances higher than 50cm.

Figure 3. A Smart Pebble. On its surface is possible to notice the hole housing the transponder

Figure 4. A moment of the localization operations

While this application is interesting because sea is probably the most complex environment for the underwater use RFID, this technology has also been employed several times for the study of sediment transportation in rivers [13-14].

All these solutions are based on the use of Low Frequency technology. 125kHz or 134.2kHz transponders are introduced inside pebbles that act as tracers in the same way as the marine application.

Anyway, differences occur in the way transponders are detected. In some applications, a reader carried by hand is employed: this means that in most cases the reader is kept outside water and used as a sort of metal detector along portions of the river where the depth is very low. Other interesting solutions are based on the deployment of an array of antennas directly on the river bed. In this case, the tagged pebbles are detected only when they pass over one of the antennas.

5. Future applications

The systems described in the previous sections represent a good starting point for the development of many other possible applications, in the same applicative fields but also in totally new ones.

Starting from the animal tracking application, the extension of this solution to other scenarios is limited mainly by the reading range, which forces the fish to come close to the reader antenna to be identified. Anyway, the chance to track animals also under water suggests a possible use of RFID technology also in the sector of fish breeding. In this case, the use of such a solution could be used to trace the production process and to guarantee the quality of the final product. On the opposite side, the use of RFID technology to trace the movements of wild fishes is notably more difficult. The RFID reading range makes the possibility to trace fishes in the sea (or even in a lake) virtually impossible because the chances that a fish will come close to some antenna positioned elsewhere are close to zero. On the other hand RFID could be used to monitor the movements of fishes along a river. In this case, antenna arrays could be structured as a sort of RFID barrier in locations where the river depth is low enough to allow the detection of every transponder passing over it. In this case, such a system could be for example useful to study the migration processes of fishes like salmons.

The technique set up for the pipeline monitoring could be easily extended to other typologies of industrial monitoring. In particular, it could be applied to monitor the state of harbour infrastructures, ship hulls, oil platforms and all the other offshore industrial plants. In all these scenarios, RFID could be useful to keep trace of the maintenance interventions performed in specific locations. The operators could use RFID transponders as a sort of electronic note where the state of the site could be read and then updated every time that any sort of intervention is performed.

The underwater navigation application could be a good starting point to develop applications where RFID is used to manage the movements of boats inside the harbours. In this case, RFID transponders could be used as a sort of electronic trail, with a reader positioned directly on the boat analysing the information stored on them and using it to move inside the harbour. On the other side, it could be possible to deploy transponders directly on the boat, and to use them as a sort of electronic license plate. This could allow the boat to be automatically identified by a reader positioned on the pier without the direct intervention of a harbour operator.

The field of environmental monitoring probably opens the way to the widest range of possible applications. Together with the geological applications concerning the sediments tracking, RFID could also be useful for the monitoring of biological activities both is rivers and in the sea. The application concerning the tracking of pebbles has in fact suggested a possible extension for this technique. The pebbles recovered at the end of the experimentation presented a lot of organic sediments left on them: this fact suggests then their possible use also as probes to analyse the impact of pollution on the biological activity of the portion of littoral under study. This technique could also be extended to be employed in other scenarios where sediments tracking is required: a similar system could be for example deployed in the city of Venice to monitor the condition of the canals. In general, such a solution could be used in those water environments where the dynamics are slow enough to keep the tracers in an area small enough be manually scanned using a reader. In this sense, such a system could be used for example to analyse in detail the dynamics of a glacier.

Together with these possible applications, deriving from the existing systems, other possible solutions could be studied every time that an under water monitoring or tracking system is required.

6. Conclusions

In this chapter the chance to use RFID technology for systems operating under water has been analysed.

The composition of salt and fresh water has been described, together with the influence that the salinity has on the conductivity of water and then on key parameters like water attenuation and penetration depth. The value of this second parameter has been calculated for the standard RFID systems: the results show that only at Low Frequencies it's possible to develop solutions where a long reading range is required, both for salt and fresh water. Anyway, moving at higher frequencies, while for fresh water the chances to set up efficient solutions are still high, especially for short range applications, for salt water RFID becomes virtually unusable.

However, the chance to use the lower frequency bands has led to the development of some applications that use RFID technology for specific purposes, both in marine and in fresh water environments. These applications range from animal tracking solutions to environmental monitoring systems, and represent a good starting point for a wider diffusion of this technology even in a sector traditionally precluded to technologies relying on electromagnetic fields for their functioning.

Author details

Giuliano Benelli and Alessandro Pozzebon

University of Siena, Department of Information Engineering, Siena, Italy

References

[1] Edwards J. Undersea with RFID, RFID Journal, http://www.rfidjournal.com, 2012

[2] Butler L. Underwater Radio Communication. Amateur Radio, 1987.

[3] Finkenzeller K. RFID Handbook: Fundamentals and applications in contactless smart cards and identification, Wiley and Sons, Chichester, UK, 2003.

[4] Goulbourne A. HF antenna design notes, technical application report, Texas Instruments, Radio Frequency Identification Systems, Tech. Rep. 11-08-26-003, 2003.

[5] Development and Implementation of RFID Technology, In-Tech, Wien, Austria, 2009.

[6] Intermec RFID System Manual, Intermec Technologies Corporation, 2005.

[7] Realizing the promise of RFID: insights from early adopters and the future potential, EAI Technologies, 2005.

[8] Bacheldor B. A Fish Tale, RFID Journal, http://www.rfidjournal.com, 2011

[9] Collins J. Taking RFID at new depths, RFID Journal, http://www.rfidjournal.com, 2006

[10] Harasti T.J., Howell J.E. and Hertel W. M. Underwater RFID Arrangement for Optimizing Underwater Operations. United States Patent Application Publication, 2011

[11] Benelli G., Pozzebon A., Raguseo G., Bertoni D., Sarti G. An RFID based system for the underwater tracking of pebbles on artificial coarse beaches. Proceedings of SENSORCOMM 2009, The Third International Conference on Sensor Technologies and Applications, pp. 294-299, Athens, Greece, 18-23 June 2009

[12] Benelli G., Pozzebon A., Bertoni D., Sarti G. An analysis on the use of LF RFID for the tracking of different typologies of pebbles on beaches". Proceedings of IEEE RFID/TA 2011, IEEE International Conference on RFID-Technologies and Applications, Sitges, Spain, 15-16 September 2011

[13] Schneider J., Hegglin R, Meier S., Turowski J.M., Nitsche M. and Rickenmann D. Studying sediment transport in mountain rivers by mobile and stationary RFID antennas. In River Flow 2010, Dittrich, Koll, Aberle & Geisenhainer (eds), Bundesanstalt für Wasserbau ISBN 978-3-939230-00-7, p. 1723-1730, 2010/09/01

[14] Lamarre H., MacVicar B. and Roy A.G. Using Passive Integrated Transponder (PIT) tags to investigate sediment transport in gravel-bed rivers. In Journal of Sedimentary Research, v.25, 736-741, 2005

Permissions

The contributors of this book come from diverse backgrounds, making this book a truly international effort. This book will bring forth new frontiers with its revolutionizing research information and detailed analysis of the nascent developments around the world.

We would like to thank Mamun Bin Ibne Reaz, for lending his expertise to make the book truly unique. He has played a crucial role in the development of this book. Without his invaluable contribution this book wouldn't have been possible. He has made vital efforts to compile up to date information on the varied aspects of this subject to make this book a valuable addition to the collection of many professionals and students.

This book was conceptualized with the vision of imparting up-to-date information and advanced data in this field. To ensure the same, a matchless editorial board was set up. Every individual on the board went through rigorous rounds of assessment to prove their worth. After which they invested a large part of their time researching and compiling the most relevant data for our readers. Conferences and sessions were held from time to time between the editorial board and the contributing authors to present the data in the most comprehensible form. The editorial team has worked tirelessly to provide valuable and valid information to help people across the globe.

Every chapter published in this book has been scrutinized by our experts. Their significance has been extensively debated. The topics covered herein carry significant findings which will fuel the growth of the discipline. They may even be implemented as practical applications or may be referred to as a beginning point for another development. Chapters in this book were first published by InTech; hereby published with permission under the Creative Commons Attribution License or equivalent.

The editorial board has been involved in producing this book since its inception. They have spent rigorous hours researching and exploring the diverse topics which have resulted in the successful publishing of this book. They have passed on their knowledge of decades through this book. To expedite this challenging task, the publisher supported the team at every step. A small team of assistant editors was also appointed to further simplify the editing procedure and attain best results for the readers.

Our editorial team has been hand-picked from every corner of the world. Their multi-ethnicity adds dynamic inputs to the discussions which result in innovative outcomes. These outcomes are then further discussed with the researchers and contributors who give their valuable feedback and opinion regarding the same. The feedback is then collaborated with the researches and they are edited in a comprehensive manner to aid the understanding of the subject.

Apart from the editorial board, the designing team has also invested a significant amount of their time in understanding the subject and creating the most relevant covers. They scrutinized every image to scout for the most suitable representation of the subject and create an appropriate cover for the book.

The publishing team has been involved in this book since its early stages. They were actively engaged in every process, be it collecting the data, connecting with the contributors or procuring relevant information. The team has been an ardent support to the editorial, designing and production team. Their endless efforts to recruit the best for this project, has resulted in the accomplishment of this book. They are a veteran in the field of academics and their pool of knowledge is as vast as their experience in printing. Their expertise and guidance has proved useful at every step. Their uncompromising quality standards have made this book an exceptional effort. Their encouragement from time to time has been an inspiration for everyone.

The publisher and the editorial board hope that this book will prove to be a valuable piece of knowledge for researchers, students, practitioners and scholars across the globe.

List of Contributors

Ela Sibel Bayrak Meydanoğlu
Turkish-German University, Faculty of Economics and Administrative Sciences, Department of Business Administration, İstanbul, Turkey

Müge Klein
Marmara University, Faculty of Administrative Sciences, Department of Business Informatics, İstanbul, Turkey

M. T. de Melo and B. G. M. de Oliveira
Federal University of Pernambuco, PE, Recife, Brazil

Ignacio Llamas-Garro and Moises Espinosa-Espinosa
Centre Tecnològic de Telecomunicacions de Catalunya (CTTC), Communications Subsystems, Castelldefels, Spain

Janne Häkli, Kaarle Jaakkola and Kaj Nummila
Sensing and Wireless Devices, VTT Technical Research Centre of Finland, Espoo, Finland

Antti Sirkka and Ville Puntanen
Tieto Oyj, Tampere, Finland

Yu-Cheng Lin, Yu-Chih Su, Nan-Hai Lo, Weng-Fong Cheung and Yen-Pei Chen
National Taipei University of Technology, Civil Engineering, Taiwan

Elena de la Guía, María D. Lozano and Víctor M.R. Penichet
University of Castilla-La Mancha, Spain

Alberto Regattieri and Giulia Santarelli
DIN – Department of Industrial Engineering, University of Bologna, Bologna, Italy

Juraj Vaculík, Peter Kolarovszki and Jiří Tengler
University of Zilina, Faculty of Operation and Economics of Transport and Communications, Department of Communications, Slovakia

Giuliano Benelli and Alessandro Pozzebon
University of Siena, Department of Information Engineering, Siena, Italy

9 781632 404350